Reimagining Rural Transformation

This book examines the effects of the pattern of growing integration between the rural and urban economies in India. Drawing on in-depth surveys conducted in villages in north India, it examines the rural agricultural economy's transformation, productivity, technology deployment, and social relations over a period of seven years. The book focuses on the socially embedded nature of the dynamics of transformation, weaving analysis around the axis of land, caste, and gender. It also identifies policy gaps and recommends steps for a sustainable and inclusive rural transformation in the Global South.

An important contribution to the study of India's economic and social landscape, this book will be useful for scholars of agriculture, sociology, economics, political science, development studies, and South Asian studies. It will also be of interest to policymakers and journalists interested in rural development, migration, employment, agriculture, and demography.

Prashant K. Trivedi, Associate Professor, Giri Institute of Development Studies, Lucknow, India.

I0894774

Reimagining Rural Transformation

Market Dynamics and Social Inequalities in North India

Prashant K. Trivedi

Routledge
Taylor & Francis Group
LONDON AND NEW YORK

First published 2025
by Routledge
4 Park Square, Milton Park, Abingdon, Oxon OX14 4RN

and by Routledge
605 Third Avenue, New York, NY 10158

Routledge is an imprint of the Taylor & Francis Group, an informa business

© 2025 Prashant K. Trivedi

The right of Prashant K. Trivedi to be identified as author of this work has been asserted in accordance with sections 77 and 78 of the Copyright, Designs and Patents Act 1988.

All rights reserved. No part of this book may be reprinted or reproduced or utilised in any form or by any electronic, mechanical, or other means, now known or hereafter invented, including photocopying and recording, or in any information storage or retrieval system, without permission in writing from the publishers.

Trademark notice: Product or corporate names may be trademarks or registered trademarks, and are used only for identification and explanation without intent to infringe.

British Library Cataloguing-in-Publication Data
A catalogue record for this book is available from the British Library

ISBN: 978-1-032-96145-3 (hbk)
ISBN: 978-1-032-96151-4 (pbk)
ISBN: 978-1-003-58824-5 (ebk)

DOI: 10.4324/9781003588245

Typeset in Sabon
by Deanta Global Publishing Services, Chennai, India

In loving memory of my father Sri Dayal Trivedi

Contents

	List of Tables	*viii*
	Acknowledgements	*x*
1	Introduction: The Paradox called Rural India	1
2	The Transforming Rural Landscape: The Story of Four Villages	28
3	Inequalities of Opportunities and Outcomes	51
4	Gender, Capital, and Occupational Change	76
5	Market, Caste, and Land	95
6	Agriculture, Productivity, and Technology	118
7	Social Inequalities, Market Dynamics, and the State	139
	References	*175*
	Index	*183*

Tables

1.1 Social Group-Wise Distribution of Sample Households Surveyed both in 2013 and 2020 24
1.2 Religious Group-wise Distribution of Sample Households Surveyed both in 2013 and 2020 24
3.1 Population Distribution in Different Age Groups: 2013 and 2020 54
3.2 Population Distribution in Different Social Groups: 2013 and 2020 56
3.3 Social Group-wise Annual Growth Rate of Population between 2013 and 2020 57
3.4 Age-group-wise Annual Growth Rate of the Population between 2013 and 2020 57
3.5 Population Distribution by Activity Status: 2013 and 2020 58
3.6 Population Distribution by Educational Status: 2013 and 2020 63
3.7 Different Social Groups by Relative Distribution of Household Durable Assets, 2013 and 2020 69
3.8 Different Social Groups by Relative Distribution of Financial Assets, 2013 and 2020 72
4.1 Population Distribution by Primary Occupation: 2013 and 2020 81
4.2 Population Distribution by Secondary Occupation: 2013 and 2020 83
4.3 Distribution of Observations of Secondary Occupation as Per Primary Occupation, 2013 and 2020 85
4.4 Occupational Distribution of Male, 2013 and 2020 87
4.5 Occupational Distribution of Female, 2013 and 2020 88
5.1 Distribution of Households by Land Category (Acres) in Different Villages – 2013 and 2020 100
5.2 Distribution of Households by Social Group and Land Holding in Senapur – 2013 and 2020 102
5.3 Distribution of Households by Social Group and Land Holdings in Seemli Village – 2013 and 2020 105

5.4 Distribution of Households by Social Group and Land Holdings in Pandari Village – 2013 and 2020 107
5.5 Distribution of Households by Social Group and Land Holdings in Gohanakallan Village – 2013 and 2020 108
5.6 Distribution of Operated Area by Social Groups and Villages – 2013 and 2020 110
5.7 Percentage of Leased-in Land in Total Operated Area – 2013 and 2020 111
5.8 Total Agricultural Land Area (acre) Purchased and Sold by Households between 2013 and 2020 113
5.9 Reasons for Sale of Land in Different Villages (Percentage) 114
6.1 Cropping Pattern (Land Sown in Acre) in Four Villages: 2013 and 2020 124
6.2 Yield (Quintal Per Acre) of Different Crops in Four Villages: 2013 and 2020 129

Acknowledgements

The driving force behind this book was Prof Manoranjan Mohanty, who consistently encouraged me to publish my work in the form of a monograph. Although Prof Mohanty was formally my senior colleague at the Council for Social Development, New Delhi, I consider myself fortunate to have him as my teacher. He initiated me into several new areas of social science research, exposed me to a wealth of new ideas and people, and provided me with numerous opportunities. One of Prof Mohanty's best qualities is that he finds positive value in every single human being who comes across him. He is democratic to the core and while working with him, people would not realise that they were faced with an authority on political theory.

I also want to express my gratitude to Prof K B Saxena, who provided valuable guidance in the field of land and rural development studies. He is the finest human being and compassionate scholar I have ever met in my life. He defies the stereotypical image of a bureaucrat.

I take this opportunity to thank Prof Nadeem Hasnain, who has been my teacher and comrade since my early academic days. His unwavering support has been instrumental in propelling my journey in social science research.

Prof Surinder Kumar is another teacher and mentor of mine, who also happened to be the Director of the Institute where I worked. Prof Kumar taught me Political Economy theory and methodology. Despite holding degrees in Sociology, after spending several years in his close company, I no longer feel that Economics is an alien discipline.

The guidance of my mentors greatly influenced the knowledge and insights that went into preparing this monograph. While I may not have been their formal student, Prof Mohanty, Prof Saxena, Prof Kumar, and Prof Hasnain bestowed affection upon me, nurturing a supportive student–teacher relationship.

It was in 2012 and 2013 when the idea of instituting a series of village surveys took shape in the Giri Institute of Development Studies (GIDS), Lucknow. Prof A K Singh, the then Director of the Institute, initiated a baseline survey that would serve as a benchmark for future studies. He chose villages that had been studied previously by other scholars in different parts of the State and deployed a team of faculty members to conduct a baseline survey in 2013. ICSSR, New Delhi, had funded that survey.

After the completion of the term of Prof Singh, most of the work on the baseline survey was carried out during Prof Surinder Kumar's tenure as Director of the Institute. Prof Kumar encouraged me to undertake an in-depth study of the rural transformation in 2017–2018. For some reasons, the village survey was finally completed in 2020.

My colleague at the Institute, Dr Shivakar Tiwari, made a valuable contribution to data analysis. He painstakingly matched the two data sets and produced flawless results by investing a huge amount of time and energy in this challenging job. The field team consisting of Shri Munna Lal Yadav, Dr Ashok Maurya, Mr Dinesh Yadav, Mr Vinod Kumar Verma, and Mr Ram Bilas collected a massive amount of data from the sample villages and households. Ms Jai Laxmi, Mr Abhishek Banerjee, and Mr Shubham Verma entered the data into the computer.

I am also thankful to the GIDS accounts and administrative team comprising Ms Geeta Bisht, Mr Krishan Kumar, and Mr Ashok Kumar for their support. The library team of GIDS, including Ms Anjali Srivastava and Ms Rita Sinha, has always been supportive.

The study was funded by the Indian Council for Social Science Research (ICSSR), New Delhi. I am particularly thankful to Ms Revathy Vishwanath for her continuous support.

My spouse, Abha, is the cornerstone of my life; without her encouragement, this book would not exist. My daughters, Veera and Pakhi, bring me immense joy. My mother Vibha Trivedi, instilled compassion in me, enabling me to appreaciate others' perspectives. This book is as much theirs as it is mine.

Prashant K. Trivedi
14 June 2024

1 Introduction

The Paradox called Rural India

Social scientists and policymakers have faced a paradox in the post-pandemic rural landscape that has become increasingly challenging to untangle. Until recently, the story of the rural was on dotted lines. Workers were migrating from rural to urban areas and from agricultural occupation to non-agricultural ones, a structural transformation further confirmed by agriculture's diminishing importance in the rural economy. The first shocker came from the official data showing that workers were returning to agriculture, a reversal unprecedented in the history of India and other countries. Hopes were pinned on the recovery of economic activities in towns after the temporary pandemic disruption. But people had to come out of denial mode as the surveys reiterated the "retrogression" alongside the official declaration of leaving the pandemic behind and the economy picking up and surpassing pre-pandemic expectations (Damodaran, 2024a, Himanshu, 2024, Satheesha and Bailwal, 2024). If the reversal of the inter-sectoral movement of workers was insufficient, the falling female labour force participation rate (LFPR) also witnessed a turnaround. This change was more pronounced in rural areas. We were told that the withdrawal of women from the workforce reflected the growing prosperity of rural households, confirming the famous U-curve hypothesis (This issue has been discussed in some detail in Chapter 4). Nobody came forward yet to claim that the reversal of falling female LFPR reflected the onward march on the U-curve, except the official position attributed to government schemes freeing women's time from the drudgery of fetching water or cooking fuel. Then came the "paradox of stagnant rural" wages, revealing the diverging paths between headline GDP growth numbers and real rural wages (Damodaran, 2024b). The initial speculative remark would be that the agriculture sector, which employs many workers, might not be doing well, but to everyone's surprise, that was not the case. On the contrary, the sector performed better in the current decade than in the previous decade (Damodaran, 2024c; Santolo et al., 2024; Himanshu, 2024). While social scientists have aptly captured distress signs emanating from the rural economy, in the backdrop of the loss of agricultural dynamism despite shifting cropping patterns towards commercial crops, the emergence

DOI: 10.4324/9781003588245-1

of the non-farm sector in the driving seat, and increasing market penetration in the rural economy (Himanshu, 2024), the failure to understand the diverging trends has only added to the confusion. Writings from a macro view of the economy have offered plausible explanations in terms of capital-intensive and labour-displacing growth, employment crisis in non-farm sectors spilling over to agriculture leading to a race to the bottom, consequently depressing wages. These works consider women's increasing participation in low-productivity sectors a crisis-driven phenomenon (Damodaran, 2024b, Goel, 2024). However, the problem with macro data and large-scale official surveys is that they miss out on changing production and social relations. They would be unable to understand how the burden of a crisis is transferred to a weaker party in an equation within or outside the home. These nuances have been effectively captured by village studies, which have recently seen a resurgence. Singh and Gaurav's (2024) work is an excellent example of this growing body of literature. Similarly, Das et al. (2024) shed light on the changing composition of household income over two decades in a village. Kumar and Raut (2024) and Arora (2024) are other noteworthy contributions to this tradition. These single-village studies offer in-depth insights into micro-processes at work, but the absence of a comparative framework severely handicaps their explanatory power. They have no opportunity to compare one village's transformation with another in the context of their unique history, sociology, and political economy to pinpoint the differences that matter. Further, large-scale surveys and village studies miss the elephant in the room. These works need to explicitly recognise that the rural transformation India and much of the Global South were experiencing in contemporary times occurred under neoliberal globalisation, which was marked by the state's increasing tilt towards capital, the market acquiring a pre-eminent status in public policies on development and widening social inequalities. The scale of change that has come about in a short period also entails revisiting the fundamental propositions of transformation.

The term "rural transformation" was popularised in Indian academia by Kurien (1983) in his pioneering study. Later, several researchers deployed this concept in their studies on rural India, but theoretically, this notion was rarely adequately addressed. Before we proceed to the intricacies of the debate on "transformation" and its relevance to the present study, it should be noted that much before social sciences, biological sciences had used the term "transformation" to denote a change in the life form.

Social sciences have witnessed a powerful "transformation" versus "transition" debate. In this discourse, the term "transformation" meant a qualitative change in production relations, whereas "transition" referred to quantitative changes. The basic premise underlying this debate was that quantitative changes were incremental, continuous, and consistent, resulting in a qualitative change after reaching a threshold. In political economy literature, this debate was primarily concerned with the "transition within a mode of production" and "transformation from one mode of production to another". Moreover, the term transformation signified a sweeping change encompassing social structures, political institutions, economic relations, and cultural ethos. They were supposed to change in tandem with each other (Kumar, 2021). This precisely meant that society's sociology, economics, and politics evolved together and replaced institutions and ideas belonging to an older society. And this last point was the trickiest part, at least from the point of view of the present study. It needed to investigate whether and to what extent emerging relations, institutions, and ideas were compatible. And whether they were changing simultaneously or was there any lag.

In macroeconomics, transformation refers to a movement from an agrarian economy to an industrial economy and further to a service-sector-led economy. It was not just a switch from a primary sector-dominated economy to one where secondary or tertiary sectors occupied the preeminent position in terms of share in the economy, but it accompanied a transformation of the occupational pattern (Adams, 1994). These Eurocentric theories took a rather deterministic and linear view of change. This argument was echoed in Stiglitz (2001, p. vii) when he argued that the "transformation of European civilisation is analogous to the transformation confronting developing countries around the world today". Indian reality was complex, and a clear break from the past might not be possible. However, no society has wholly shunned the baggage of pre-capitalist relations, but it was more so in the case of southern countries.

The most potent conceptualisation of "transformation" could be found in Polanyi (1944) in his magnum opus *The Great Transformation*, which describes the changes experienced by Europe during industrialisation in the nineteenth and early twentieth centuries along with the massive social, economic, and cultural changes that accompanied this period. He underlines the crucial link between politics and economy (Stiglitz, 2001).

Polanyi (1944) gave a profound concept of embeddedness to debunk theories of the autonomous economy and self-correcting markets. He forcefully argued that markets have permanently been embedded in the economy and that the economy was deeply implanted in social relations, political institutions, and culture. Contrary to the view of mainstream economic thought, Polanyi refused to accept that capitalism has been successful in disembedding the economy. For him, it would require humans and nature to be converted into a pure commodity whose prices would be governed by markets (Block, 2001).

Polanyi worked on macro-level transformation, but the insights one drew from his writings were that any transformation study should never see economics, politics, sociology, and culture as autonomous from each other. Instead, their interconnectedness was the most interesting facet to be studied. Other political economists also saw social relations, the economy, and power structures as aspects of the same whole, shaping each other with causal relationships working both ways.

Several contemporary economists further took Polanyi's work ahead. Adams (1994, p. 332) described "transformation" in these words:

> Since roughly 1800, the world's peoples have been passing through an all-encompassing and unparalleled systemic transformation. This transformation has been driven by individual and collective actors involved in clashing political and intellectual movements of sweeping importance. In slightly different language: the great transformation is holistic, and it involves the remaking of self and society-the conjunction of person and institution...; this process is not without meaning because, through instrumental valuing, it yields grudging progress in the amelioration of the human condition; hence, the denotiveness of change, transformation, progress.

Adams (1994, p. 343–344) further delineated:

> When change is widespread and institutions are going through rapid reconfiguration, we speak of transformation or revolution...institutional changes that fostered Europe's industrial revolution. The rise of new technologies, epitomised in the factory production system, was accompanied by the attempt to create a self-regulating market system. At the end of the twentieth century, national and world politics are still overwhelmingly driven by the social forces unleashed in autochthonous reactions to that Utopian experiment.

Another dimension of the debate on transformation was related to the negotiation of emerging economic, political, and social forces with pre-existing relations. Caste figured prominently in this discourse. The persisting relevance of caste in the contemporary economic, social, and political milieu provided a perfect vantage point for this debate that questioned the earlier understanding that the penetration of market forces would replace pre-existing relations often premised on caste. It was believed that the market did not care about a person's identity and that all actors had equal weight in carrying out a transaction. It put market forces at odds with primordial identities; both engaged in a battle of existence (Bhaduri, 2010; Jerath, 2013). The same understanding was reflected in the developmental strategy adopted by the Indian establishment that facilitated the penetration of market forces in the countryside, believing that it would transform the land, labour, and commodity relationships in rural areas from status to contract.

Against this "replacement" view, another perspective saw "modification" and "appropriation" as more appropriate categories to define processes underway in southern countries. They saw market forces modifying caste instead of eradicating it. For instance, caste might not remain the only determinant of occupation, but it reflects the capacities required for occupational diversification. Similarly, land-caste congruence has been modified, but not annihilated, under the impact of the massive land reforms programme. Social science literature has noted the tendency of market forces worldwide to preserve the pre-existing structure that suited its expansion. In many cases, these structures were modified to make them suitable.

Agrarian Change, Non-Farm Economy, and Migration

The thick overlap between agriculture and rural areas was reflected in social sciences investigating rural transformation through the prism of agrarian change and its linkages with the rural non-farm sector and the entire social formation, specifically modern industries. The idea was to probe the interaction between emerging capitalism, the supposedly non-capitalist agrarian sector, and the non-farm economy between the two major sectors. These research studies were based on the a priori assumption of the ultimate domination of capitalism and the subsequent articulation of agriculture with it (Breman and Mundle, 1991).

These crucial theoretical and empirical exercises set out with the assessment of the production relations in agriculture. For instance, semi-feudalism, capitalism, and dual-mode were frequently referred to as nomenclature in this deliberation, which was known as the "mode of production debate" (Thorner, 1982). Secondly, social scientists invested in this discourse presented their assessment of the characteristics of the transition "path" that agriculture was likely to traverse during the journey towards capitalism. These "paths" were generally deduced from the experiences of the countries where capitalism had already overwhelmed the agrarian sector, for instance, the American Path, Prussian Path, Japanese Path, etc. (Byres, 1991).

All these paths would consist of the dynamics of the internal class composition of the agrarian sector under transformation. One of them, the powerful concept of differentiation, was particularly influential in debates in India. Premised on English experiences, it believed in the emergence of a class of capitalist farmers that would invest in hired labour-based agriculture and a larger class of wage labourers under the polarising effect of capitalism on the relatively homogenous peasantry (Patnaik, 2007). However, the possibility of full articulation of agriculture within the capitalist system without the replacement of the peasantry by a wage labour–capital relationship was not ruled out (Byres, 1991).

Beyond the internal dynamics of the agrarian sector, its linkages with the other sectors of the economy were found to be more relevant for rural transformation. In Marx, three crucial linkages based on the flow of capital, labour, and commodity were identified. First, the surpluses generated by agriculture were accumulated as industrial capital, specifically in the initial stages of industrialisation. Second, the surplus labour released from agriculture was absorbed as wage labour in industries. Third, with a huge population dependent on it, the farm economy and households engaged in it provided a "home market" for industrial goods (Breman and Mundle, 1991).

The three linkages continued to inform research on rural transformation in the 20th century, however, with shifting relative emphasis on one or the other. The agricultural sector's function as a home market came into the limelight initially, but in more populated Asian countries, the labour link between the primary and secondary sectors proved to be crucial. This link had the potential to trigger a virtuous cycle between agriculture and industry, specifically the rural non-farm industry, the sector that had shown the highest tenacity to absorb surplus labour, but the main question was whether the agrarian sector was offloading labour under duress or whether it was a willing movement towards a higher productivity sector (Breman and Mundle, 1991; Harris, 1991).

The oft-discussed Mellor (1968) model of the "virtuous spiral" between food production and employment generation could give a clue of the initial thinking on the subject. The model was based on the premise that the deployment of cost-saving technology and increased food production, resulting in higher contributions to national income, particularly from large farmers, would eventually boost demand for urban industrial goods and services. The virtuous cycle was completed with rising employment in industry and demand for food grains. The Mellor model was criticised for assuming that agricultural growth would necessarily lead to rural employment diversification, pointing out that it was necessary but not sufficient. There was no evidence to prove that movement out of agriculture would inevitably lead workers to modern industry. In fact, a large chunk of them had been absorbed in rural non-farm enterprises (RNFs) (Harris, 1991; Breman and Mundle, 1991).

Another pertinent question from the perspective of the present study was about the driving seat. Whether agricultural growth pulled the diversification or the big industry drove it. And in the second case scenario, whether the primary sector was a drag on it (Harris, 1991). Agriculture growth-led diversification was qualitatively different from the one led by industry. Recently, agricultural growth in many regions of India has been slower than in the fast-moving sectors of the economy. But industry-led diversification did not mean a simple difference between the growth rates of the sectors. It might entail a situation where agricultural workers find industry a more attractive option. It also meant a condition of involution where the carrying capacity of agriculture was exhausted and the sector could not provide adequate income to people engaged in it. The second condition was qualitatively different from the first (Harris, 1991). This point is discussed in Chapter 6. Suffice it to add that both processes reinforced each other. The pressure on per capita income from agriculture and the availability of jobs in urban industries have been conjoined.

Social science evidence on the movement from agriculture to non-agriculture has been mixed. In the 1970s and 1980s, when a major chunk of surplus agricultural labour was absorbed by rural non-farm enterprises (RNFs), the sector was hailed as a major employment provider. However, some works found this movement to be of an involutionary type, distress led the migration to a sector of low productivity. Around the same time, some other studies found this movement to be prosperity-led. These studies challenged the homogenisation of RNFs, empirically documenting broadly two types of enterprises on the basis of labour characteristics, level of technology, and capital infusion. But broadly, South Asia was home to RNFs that used elementary technology and family labour, particularly women's labour, and were driven by a seasonal oversupply of labour rather than demand for goods in the market. In north India, where the present study was located, RNFs were essentially subsistence activities (Harris, 1991; Mukhopadhyay and Lim, 1985; Islam, 1987; Chakrabarti and Kundu, 2009).

Besides spatial variations, longitudinal diversities betrayed a homogenous narrative about RNFs in India. After the 1990s, the subject of RNFs almost went off the academic radar, which also reflected the diminishing importance of the sector in the face of intensifying market intervention. Alongside this, a particular change related to RNFs' relationship with agriculture has been observed. For instance, the rich peasantry-focused Mellor's model was based on food and employment, while later development in this theory underscored that egalitarian growth in agriculture was conducive to the growth of RNFs. In both cases, a thick relationship between the two counted. There were certain pressures on the mutually reinforcing equation, for instance, movement away from food grain (Chakrabarti and Kundu, 2009), but an unprecedented flow of factory-manufactured goods in a couple of decades changed the fate of RNFs drastically.

The emerging RNF sector was markedly different from its incarnation in the 1980s in its relationship with agriculture and the overall market economy. While the earlier RNFs had a close relationship with agriculture at both input and output ends. They provided services to the cultivation, and processed and traded agricultural output. By this time, the Jajmani-era service providers were already replaced by technicians dealing with modern machines, but processing and trade were largely limited to the commodities produced locally. In a nutshell, the entire RNF ecosystem was woven around agriculture. So close was this intertwining that RNF frequently reflected the growth patterns and technical changes in agriculture. They also played the role of an intermediary between agriculture and the market economy, mostly in processing, trade, and certain technical services.

On the other hand, the new RNF was largely an extension of the market economy, largely cut off from the agriculture sector. These were mostly industrial goods retailers and service providers that essentially brought erstwhile urban experience to the rural neighbourhood. However, agro-processors and traders have not vanished, but they appeared eclipsed by the exponential rise of trade in consumer goods, transportation, communication, infotainment, educational, and health services. The site of RNF has witnessed a colossal decoupling of agriculture and rural.

However, the dynamics of the agrarian economy did impact RNF, not just in the agro-processing and agri-trade sector but also in the industrial goods and urbane consumer segment. The crucial difference was that the latter was no longer dependent on the former. Agricultural incomes continued to boost the commodity market, but with the opening of a new stream of cash flow, however narrow, discretionary expenses often contingent on urban incomes.

The works that cautioned against the disjunction of farm and non-farm sectors due to crop diversification (Chakrabarti and Kundu, 2009) did not anticipate a massive transformation of the rural economy by the capital. The rich farmer-based or egalitarian growth-led models have ceased to be relevant in the new milieu where capital calls the shots. However, the rural economy was hardly isolated from the market economy; whatever autonomy it had has been overwhelmed by the capitalist market. The link between agriculture and RNF was crucial in sustaining the limited self-sufficiency the rural economy had, but capital has been dealing separately with different segments of this economy. We have discussed in Chapter 6 the way capital has overtaken some parts of the agrarian economy. Similarly, RNF has also been facing replacement and articulation with the capitalist economy. These processes have constrained the rural's productive capacities, converting them into cheap industrial goods and services consumers.

The second job that the capital appeared to have assigned to rural areas was the reproduction and maintenance of the labour force for urban industries. Once the productive capacities of the rural areas were destroyed, replaced, or overtaken by the market economy, the ensuing imbalance was sought to be corrected by the large-scale migration of able-bodied men and women from rural to urban areas. The process of agricultural involution might have played no less a role in scaling up the migration to gigantic proportions. Thus, a clear separation between the site of production and the site of reproduction was a feature of contemporary rural transformation.

As we will see in Chapter 3, the demographic transition experienced by villages resembled that of the country, marked by an increasing share of the working-age population. It turned rural areas into a place that ensured a steady supply of labour below the reproduction cost without burdening capitalist industries. The beauty of this arrangement was that these migrant workers could be offloaded by the urban economy anytime without any sense of accountability or remorse. They formed a dispensable workforce that routinely came back home if employers in the town didn't need them. Their expendability was nowhere more visible than during the Covid-2019 lockdown in 2020. Clearly, they were not on policymakers' minds.

Economic incapacity and migration formed a vicious cycle. On the face of it, it appeared almost a natural process that workers from backward regions migrated to the areas of economic affluence, but regional inequalities have been the construction of a conscious strategy. Otherwise, why would labour have to migrate to the areas where capitalist industries were established and not the other way around? Someone could point out the a priori assumption of natural endowments, but these regions were often backward globally and within countries. The location of third-world countries in the world capitalist system and the mineral-rich regions of central India were just some examples. As we have seen in the studied villages, the capital created labour catchment areas by destroying and linking their economies with core economies. The state actively intervened in creating labour reservoirs and industrial towns (Amin, 1974; Breman, 1991). Migration was not incidental; it appeared systemic to the expansion of capital. It would be futile to question the motive of labour behind migration when economic imbalances between regions and locations were the constitutive elements of the capitalist economy. Besides economic compulsion, probing the caste question was crucial to disentangle the migration phenomenon. It has been documented widely that the anonymity of towns gave migrant workers a sense of freedom from the caste oppression of the villages.

The "Rural"

Any rural transformation study would first have to confront questions about the notion of "rural". Besides statutory towns, the Census of India defined "urban" as any place having a minimum population of 5,000, at least 75 per cent of male workers engaged in a non-agricultural occupation, and a minimum population density of 400. And all other places that did not qualify to be designated as urban were considered "rural". This definition considered the "rural" residual of "urban" (Kumar, 2014).

Since the colonial period, in social science and policy imagination, "rural" and "urban" were perceived to be dichotomous. In the vision of colonial modernity, "rural" was the other of "urban", and rural had to imitate the former to eliminate its premodern features (Kumar, 2014). In this construct, aided by "anthropology, the discipline invented to study the colonised (Cohn, 1987), and premodern 'stagnant' societies against sociology for the study of dynamic modern societies, rural was laden with tradition and backwardness, whereas urban symbolised progress and development (Kumar, 2014; Singh, 2016). Rural was still goaded to become urban to overcome its shortcomings. No wonder 'urbanisation' was considered one of the most important yardsticks for development in third-world countries" (Kumar, 2014).

Studies have questioned the coalescence of "village" with "rural", pointing out that historically the former was a colonial administrative construct. This view saw the possibility of rural beyond villages. In the later period, "rural" came to be identified with "village", especially in the Ganga–Yamuna basin areas (Kumar, 2014). Further, when it came to villages, leaders of India's National Movement had different perceptions about them; Nehru and Ambedkar saw it as a site of backwardness and oppression, respectively; it was Gandhi's identification of "village" with the purity that became the nation's default imagination of a harmonious village republic. This "rurality" was premised on a close correspondence with agriculture as a core economic activity, the most significant source of employment, and a way of life (Jodhka, 2002, Singh, 2016).

The emerging rurality carried forward some elements from the previous sense of the term but massively restructured many other aspects. For instance, agriculture might still keep most people engaged, but it was neither considered desirable nor the most important source of income for "farming" households. A massive movement from agriculture to non-agricultural sources of income was perceptible, but people still held on to their agricultural land while quitting the sector. And this occupational diversification has not been uniform for all castes and social groups. The faster pace of SCs leaving agriculture has been noted by many studies (Jodhka, 2016; Singh, 2016).

Another remarkable feature of the emerging rurality was the disintegration of the "village community". However, this observation assumed that once there existed a unified village community that started splitting initially along caste lines. Furthermore, each household has acquired an autonomous existence free from community bonds and norms. This change also coincided with the weakening of the traditional power structure. However, the dominant families continued to be the most resource-rich and powerful, albeit differently. Power no longer flowed only from land and caste congruence in the contemporary scene. It has become an enterprise. Enormous cultural changes in rural areas and penetration of markets, goods, and services have almost demolished the rural-urban dichotomy (Jodhka, 2016; Singh, 2016).

Questioning the "Village"

The colonial narrative of the village has come under scathing scrutiny by social science research in independent India. The Metcalfian "little republics" have been subjected to enquiries from various perspectives. Interestingly, these works underlined the dynamics of a self-serving colonial construct about the Indian society emerging as common sense in the country. The colonial depiction of India was based on three major institutions where the village community found a proud place along with the caste system and joint family. The portrayal of tradition-bound India in contrast with the modern West inadvertently underlined the civilisational unity of the country. These notions were accepted uncritically by the native elite since they ensured a much-desired respectful place before colonial administration and placed them at the helm of Indian society (Jodhka, 2016).

From the beginning, drawing from the "field view", social science research has demystified the "book view" (Srinivas, 1955). The self-sustaining, autonomous village, based on interdependent relationships between families, has given way to a village located in a broader socio-economic ecosystem connected to the outer world through kinship ties, commodity markets and labour relations. Social science works have also debunked propositions regarding the static nature of village society. These works argued that a unilinear paradigm of modernity ended up portraying the rural as an "inferior other" of the urban. The narrative was premised on binaries of progressive/regressive, modern/traditional, and developed/backward. Extending these binaries, the differences between rural and urban were romanticised based on purity and morality (Singh, 2016).

However, recent social science research might come as a shock to a die-hard subscriber of this view. A series of "revisit" studies on rural transformation has resulted in findings that would demolish the myths surrounding the Indian villages. One of the recent explorations that restudied a Malwa village initially investigated by Mayer (1960) underlined the sociological importance of the village despite the disenchantment of residents with the village. The study noted that the old social, economic, and political systems were rooted in the land. It still played a crucial role in the economy and politics, but its new incarnation was land as a "market good" that provided the basis for wealth and power in a new milieu marked by heightened real estate activities (Sbriccoli, 2016).

Further, migrant Dalits cherish freedom from upper caste dominance and master–servant relationship, but urban jobs did not offer any promising future except relatively better wages. The village also provided a reference point for the success achieved elsewhere. More importantly, the integration of rural with urban has reached a point where the village internalised the city's perspective of it and began resisting encroachment (Sbriccoli, 2016).

> The village becomes part of urban imagination and is built publicly according to new values; its representation acquires new meanings for its inhabitants. The city first empties the village and then reaches to it. It encompasses and produces it as a new sociological and symbolic object, but in doing so it faces resistance in many forms.
>
> (Sbriccoli, 2016, p. 15–16)

Another remarkable revisit study investigated Bisipara, an Odia village first explored by Bailey (1959) in the 1950s. It confirmed some of the predictions made by Bailey coming true. One of them was people recognising themselves with the nation or state first rather than being primarily members of the village community. The Indian state had also sought to weaken the authority of caste in the everyday life of citizens, but success was rather limited and late and coincided with the penetration of the market in the village economy. The study could not pinpoint whether the state or the market triggered the social change in the village. But it indicated that those who were least dependent on the landed classes experienced independence in other spheres too. The study also underlined that the transformation of the relationship between landed and landless was set in motion with universal education and the arrival of mercantilism (Otten and Simpson, 2016).

A similar story of change with continuity was captured by the revisit of Palanpur, a village in western Uttar Pradesh studied numerous times during seven decades of India's independence. The latest study in the series accentuates the persisting relevance of three factors, viz demography, technology, and non-farm jobs as the drivers of change. This study argued that the way these factors intersected with each other and impacted the outcomes in the form of income, and responses of land and labour markets, among other institutions, has been changing over a period. For instance, the increasing population exerted pressure on the finite resources at the command of the village, but outmigration eased that, a process in which the availability of non-farm opportunities played a significant role. The closer integration with outside labour markets has also responded to the decreasing demand for labour in agriculture alongside mechanisation. This also affected the distribution of income and social status of different groups in the village (Himanshu et al., 2016).

Income distribution within and between rural and urban areas has been a major measure of understanding social change. An income analysis from 2014 to 2019 revealed that the income shares of the bottom of the distribution declined during the period. The analysis also found that bottom incomes experienced negative growth. And this was particularly the case with rural bottom incomes. It also signified that the bottom incomes of the country were exclusively composed of rural income, particularly SCs, STs, and small and marginal farmers. Besides revealing widening inequalities in incomes across locations and vertical distributions, the analysis showed a possible persisting trend of negative redistribution, meaning redistribution from lower to higher incomes (Sahasaranaman and Kumar, 2022).

The rising inequalities did not reflect only in income levels but also in consumption. An analysis of the National Sample Survey Office (NSSO) data from various quinquennial rounds of 1994, 2005, and 2012 revealed that non-food items increasingly contributed to widening inequalities in rural India. This finding went against the old imaginaries of the rural as the land of subsistence. Further, education has ceased to bring proportionate returns in consumption levels due to labour market functioning where rural educated people could not find appropriate jobs. This reversed the earlier role of educational achievements as an explainer of inequality in consumption levels (Pawde, 2022).

Alongside the enlarging non-food consumption, studies have highlighted booming construction activities in rural India from 1999–2000 to 2011–2012. An analysis of NSSO data revealed that during the reference period, employment in the construction industry experienced a 9.3 per cent annual compound growth rate (CAGR) against 1.5 per cent for the whole Indian economy. The analysis further demonstrated that this employment boom was largely rural, where CAGR was 12 per cent against 5 per cent of the urban. The study attributed the boom to converting *kaccha* dwellings into *pucca* houses (Mahajan and Nagaraj, 2017). This was something inconceivable for the beholders of colonial rural myths.

Another major debate concerning changing employment dynamics in rural India was about women's declining work participation in the country during the last three decades, particularly between 2004–2005 and 2009–2010. Initially, this pattern was attributed to the withdrawal of women from the workforce due to improvements in the economic status of their families. Women were not necessarily required to contribute to the family's income. Against the mainstream view based on the U-shaped curve between women's work participation and the family's economic status, other studies cautioned against hasty conclusions. They pointed out definitional anomalies in NSSO's surveys that ended up classifying non-working women principally involved in household chores, whereas they were looking for work. Once this anomaly was factored in, the fall in women's work participation would be narrowed down. Further, the mechanisation of farm operations and other technological changes in the rural economy have adversely affected the availability of jobs in villages, forcing women to withdraw to home-based activities (Siddiqui et al., 2017; Rawal and Saha, 2005).

While gender and economic development intersect to produce outcomes for women workers, caste and capitalism are also inseparable in the emerging milieu of rural areas. Studies on brick-kiln workers and sugarcane cutters have shown the persistence of bondage and exploitation where caste and capital reinforce each other cleverly. These investigations underline the coexistence of bondage with welfare schemes, with the latter subsidising the capital to keep wages low. They probe whether the motivation for upward mobility encourages Dalits to engage in consumerism and rituals, compromising their freedom to attain social prestige (Guerin et al., 2015).

Rural Transformation Studies in India

In India, social science research has taken broadly three routes to study the transformation of rural life. One of the earliest streams of probes on villages conducted ethnographic studies in various parts of the country. These investigations provided rich data on social stratification, culture, economy, and other aspects of rural society. These studies provided the basis for several rich concepts for analysing Indian society. "Sanskritisation" and "dominant caste" were two such concepts that helped sociologists appreciate the social change in Indian society. The strength of these studies lies in the long period of participant observation-based ethnographic data. But the "village" was taken as an autonomous and locally bounded entity. It had a bearing on understanding the village's interaction with the outside world through labour and commodity relations. It also constrained the capacity of these studies to explain several changes triggered by the village's location in a broader context (Athreya, 1983).

Unlike the limited perspective of the village studies tradition, the political economy investigations were always interested in the village economy's integration with the broader market economy and the accompanying transformation of institutions and relationships in the village. One strand of the political economy thoroughly probed the character of emerging agrarian relations. This group of studies investigated whether production relations in agriculture, the mainstay of the rural economy, were revealing capitalist features or were still influenced by pre-capitalist forms of domination and exploitation. To assess the character of agrarian relations, this stream of studies tried to verify the presence of capitalist farmers in Indian villages who produce primarily for the market using hired labour. Popularly known as the "mode of production in Indian agriculture debate", the effort was to comprehend the internal differentiation among the peasantry, whether it was giving way to a polarised agrarian structure with the emergence of a class of rich farmers investing in profit-oriented agriculture and a class of labourers on the other end (Patnaik, 2007). These studies conducted a threadbare investigation of every aspect of agriculture, and their insights were quite helpful in understanding the transformation that villages were experiencing.

This genre of works was found very interesting by scholars across disciplines, including economics, sociology, history, and political science. Still, policymakers did not find them equally helpful, given their abstract generalisations and complex arguments. They preferred another stream of studies that solely relied on statistical tools rather than any analytical framework (Athreya, 1983). There has been a continuous stream of rural transformation studies since the 1980s, particularly after the publication of C T Kurien's *the Dynamics of Rural Transformation* (1983).

These studies particularly appealed to policy planners for their coverage of several aspects of village life instead of remaining confined to only one of them. For instance, data on agricultural productivity, production, the spread of technology, asset ownership, employment diversification, migration, wages, access to education, and health services along gender and caste lines were collected in some studies to appreciate the multidimensional reality. They also included socio-political changes, poverty, nutrition, etc. This massive data collection exercise on time intervals helped to understand longitudinal changes (Athreya, 1983).

But social science literature critiqued these works for maintaining a distance from theoretical frameworks that incapacitated their analytical ability. Athreya (1983) argued that:

> Kurian himself decided to rely on statistical tools to understand a deliberately chosen focus rather than applying a theoretical framework. This lacuna leaves them unequipped to study relations among units as all units are considered autonomous. Besides, these studies have also been accused of ignoring power processes generated by social structure and giving excessive importance to state intervention in the form of welfare schemes in the transformation process.

Remarkably, revisit rural studies have witnessed a resurgence in the last couple of decades, underlining the importance of rural areas as a site of social science investigations (Himanshu and Rodgers, (2016). These studies collect data from the same site to build upon and question formulations put forward by scholars of previous generations. It was often believed that the contemporary revisit studies focus on change, but their earlier incarnations were primarily, if not exclusively, concerned with the stability and self-sufficiency of the villages (Jodhka, 2019). These ethnographic accounts might have been dominant in the 1950s and 1960s, but they did not go uncontested. For instance, as argued elsewhere in this work, Cohn, Opler, and the entire Senapur tradition powerfully highlighted interconnectedness with the outer world and the changing nature of rural societies.

Some recent studies have depicted rural societies as "post-agrarian" and merely a type of settlement, bereft of any distinction from other forms of settlement. As one would see in the chapters of this work, both these positions represent the other extreme of the "static" villages proposition. Agriculture continues to be at the centre of rural life as a major occupation, if not the largest source of income. The patron–client relationships, including jajmani relations and caste-occupation linkages, might not be the core constitutive elements of the village anymore, but that has not completely flattened the village as a socio-economic-political space. For instance, the village has persisted as a political unit, and traditional power sources have not withered away despite the emergence of new power sources. The power dynamics rooted in the alignment of caste and land have endured, albeit in modified forms, despite the fragmentation of the imagined "village community". Moreover, newer developments, such as Panchayatiraj, have reinforced the idea of the village as a development unit.

Martin's (2015) study further substantiates that the power struggle among castes, alongside the persisting pre-eminence of dominant groups, marked contemporary villages. These powerful interests often augment resources from the state policies, formally meant to support the marginalised groups, including Dalits. Martin's work debunks the village as a type of residential community argument by pointing out that despite SCs being elected in local bodies, rural elites were more likely to be in control of the situation. The author also underlined that SCs carefully treaded the power relations in the village, avoiding direct confrontation with the class of farmers that drew power from multiple sources seeking appropriation of state resources like village commons.

This account should not lead us to believe that India's villages were not changing. For instance, some Tamil Nadu villages have undergone changes due to local industrialisation and the effects of social policy, but that has not obliterated the importance of family farms. Interestingly, farming families in these villages have begun to prevent further division of land by encouraging additional sons to explore opportunities in the non-farm sector. Besides, the marginal and poor farming families migrate in larger numbers seeking decent living outside villages, leading to what authors call "centripetal tendency in agrarian structure". Those who stayed put in villages opted to be "pluri-active", combining various sources of income, reflecting the rising importance of non-farm income (Djurfeldt et al., 2008).

Many rural studies have posed the question of why the traditional land-owning castes did not decline in terms of landownership and political influence despite state policies aimed at curbing the inequalities. Scaria (2010) argued that the disintegration of the joint family system and consequent fragmentation of landholdings aided the downfall of landowning classes, but the spread of education and the emergence of salaried employment as the new power centre prevented this decline.

The story of the village as a type of settlement resonates with the proposition that caste is more about difference than hierarchy or power relations. Singh's (2016) account of a Rajasthan village rejected both of these formulations. In this village, Dalits striving for social mobility after achieving a degree of economic advancement faced a fierce backlash from the upper castes. These contestations resulted in caste conflicts and threatened the rural institutional framework tasked with order and stability.

The significance of caste categories was also apparent in their deployment in social science research. In another intriguing work on agriculture and rural aspirations, Tilche (2016) emphasised how a caste striving for upward mobility must actually move downward, a paradoxical shift for an outsider. This happened in the context of cultivation continuing to be remunerative, but farmers were perceived as poor, leading to a shift from agriculture to a culture of migration. Quite interestingly, Tilche's work focused on left-behind members of the caste against a rather routine practice of considering upwardly mobile individuals. Moreover, Tilche argued that the agricultural crisis in India was rather a tension between traditional values and modern aspirations that becomes evident on the site of arranged marriages.

Reimagining Rural Transformation

In social science literature, transformation signifies a movement from a primary sector-based economy to a secondary and tertiary sector-based economy (Mensah et al., 2016). It ameliorated the condition of workers by enhancing their productivity and initiating a movement from status to contract. Simultaneous changes in ideas and institutions facilitated this movement and were impacted by it. These changes transformed the entire society, consisting of both individuals and institutions. In the Indian context, gender and caste relations were crucial. It was a progressive movement towards a better life for human beings, reflecting access to material needs, reduced inequality, and social tension, resulting in social development (Kumar, 2021).

One must add that "transformation" could not be conceived as an evolutionary movement towards adopting another society's developmental and cultural parameters. Each society uniquely transformed itself. Western societies have transformed from agrarian to industrial ones before reaching service sector-based societies. Many of these core countries also claimed to have reached the stage of a knowledge-based society. The paths of transition adopted by peripheral countries have been very different. India has leapt to a service sector economy in terms of its share of GDP while retaining agriculture as the most significant source of employment. Uttar Pradesh, where the present study was located, exhibited a similar development pattern.

Further, along with the process, the transformation was relative rather than absolute in terms of forms. What a transformed society would look like might be context-specific rather than modelled around Western societies. Similarly, transformed institutions might not mimic those in core countries. This observation was particularly relevant during the transition. For instance, the transformation of agrarian relations continued to take forms entirely different from the experiences of the advanced capitalist countries.

Then, what was "transformation" in the Indian context? The suggestions mentioned above that the process and forms of transformation in India might be different from the West didn't mean that transition and transformation of relations and institutions were not taking place in India. Penetration of technology and capital, migration, and consequent diversification of employment have altered the economy of the "rural" society beyond recognition, debunking claims about its isolation from the market economy. However, the rural economy was never wholly autonomous of the broader economy, but the emerging commodity and labour relations have almost completed this integration.

The closer integration of the rural economy with the broader market economy has created a mosaic of relationships due to the interface of this process with pre-existing relations. The dynamics of integration would not have equally impacted all people in the rural areas. One of the factors influencing the outcomes might have been the location of a person in a pre-existing economy and the asset base and skill set carried forward by them. The questions of caste and gender inevitably come up in this discussion since these two axes of domination and privileges have been crucial in determining legacy and class.

Looking at it this way, rural transformation in the Indian context would consider modifying the structure of privileges and deprivation based on caste and gender. This would reflect the influence of pre-existing privileges/deprivation on emerging opportunity structure. This would mean that transformation is not just about the transition from a pre-existing opportunity structure to a new one but also the study of the dynamics of this transition. Most studies have generally considered points A and B and neglected the path between these two points, where extra-economic factors like caste and gender play a decisive role.

This lacuna in rural transformation studies stems from understanding transformation as a macro phenomenon. Underlying this premise was the assumption that the entire economy and associated structures moved in tandem at a similar pace in a particular direction. Hence, an aggregate analytical framework was preferred over a disaggregated analysis. This exercise overlooked the influence of "extra-economic factors" on economic outcomes.

Another factor limiting the explanatory power of rural transformation studies was the assumption of linear movement. This emphasis on linearity did not leave scope for discussion on any other form of movement, be it "zigzag", "back and forth", or "current and counter-current". In historically conditioned societies, like much of the world, transition taking unexpected twists and turns was not unusual, at least in the short run.

Similarly, the coexistence of different sets of production relations in southern countries sets them apart from core countries where transformation was assumed to be complete by replacing the old set of relations and institutions. Moreover, this coexistence of various modes of production might not be a temporary phase, as was presumed earlier. Recognition of this reality altered our understanding of transformation significantly. Social science evidence has underlined the coexistence of large capitalist sectors alongside petty commodity production and workers rotating between different jobs in different sectors throughout the year in India (Sanyal, 2007). Often, it has been observed that sub-sectors within sectors transform differently. The fluid nature of the emerging empirical reality of the southern countries betrayed an earlier understanding of a clear break between modes of production articulated based on Western experiences. This also rendered static social science categories irrelevant in a probe of the transformation of economic relations. Categories that could capture this movement across sectors and overlap between modes of production would be more useful in these investigations.

Underlying lacunae in Eurocentric theory to understand the emerging situation in peripheral countries didn't imply that we were discussing Indian exceptionalism. India was a market economy, and capitalism's dynamic overwhelmed the entire economy and society and didn't remain confined to one or the other sector (Namboodiripad, 1984). The non-capitalist sectors were also part of the capitalist dynamic (Sanyal, 2007). The variances mentioned above differentiated the capitalist model of the periphery from that of the core. Hence, it has been pointed out that one should not expect Western-style transformation to take place in India.

The case of India lies in the social structure and its interface with economic relations, especially the institution of property and labour relations. In concrete terms, the intersection of caste and gender with land in rural areas emerges as the analytical axis around which the story of transformation needs to be woven. Similarly, labour relations flowing from gendered and casteist property relations need to be analysed in an intersectional framework. In a typical transformation study, the investigators would probe whether and to what extent land has become a commodity, freely traded in the market (Namboodiripad, 1984). But in Indian conditions, the probe will also be conducted at the interface of caste-gender with land relations, attempting to ascertain whether it is gaining strength or weakening.

Similarly, the moot question in the investigation of labour relations would be a change from "status" to "contract". To grasp this movement would require determining whether and to what extent labour was free to choose their employer. Recent investigations have found the interlocking of the labour market with land and credit markets resurfacing. These interlockings interfere with the labour's free will to choose their employer (Trivedi, 2011). This complex set of relationships also has a close correspondence with the gender–caste equation. This will entail investigating the movement from status to contract in the context marked by the caste–gender complex.

The third question one was expected to pose relates to the diversification of sources of employment. Typical transformation studies would consider this phenomenon at the macro-level to understand the sectoral composition of the workforce at two given points. But it was not just about the movement from agriculture to non-agricultural sectors; the dynamics of this transition need to be probed. This investigation might yield insights into the role played by the land–caste–gender complex (Lerche and Shah, 2018) in shaping this journey and possibly the destination in the new opportunity structure. It needed to be examined whether diversification was accompanied by upward mobility, downward mobility, or was a horizontal transition for one or the other sections of the population under study.

The fourth aspect of transformation that needed careful attention was access to social development viz education, health, nutritious food, housing, drinking water, sanitation, social equality, political participation, and, more importantly, equality in decision-making. The caste–gender–land complex would have had the most visible impact on this aspect of rural life. Social development has always been considered a means to growth and transformation, whereas, in the newer understanding, it was regarded as an end. Looking this way, social development indicators of a society or a group were a true testimony of its transformation.

While investigating the transformation of a society and economy, it is desirable to identify factors responsible for the transformation along with impediments to this process. The penetration of capital was the most critical driver of the transformation of an economy. Investment in agriculture, allied activities, rural industries, the market for agricultural and other rural products, and increasing consumption of industrial goods in rural areas have massively transformed the "rural".

Several studies have identified technology as one of the key triggers of the transformation (Mensah et al., 2016). Technological changes acted as a double-edged sword. At one level, they decreased drudgery and increased labour productivity, but they also displaced workers. However, in Western academia, this process has often been referred to as the "release" of workers from the primary to the secondary and tertiary sectors. In a country like India, where open unemployment was high, "displacement" appeared to capture the dynamics more appropriately.

Demographic changes also contributed significantly to this process. Circular migration of workers across rural–urban settings and sectors has been historically transforming societies. This factor has gained immense momentum in the recent past, transforming societies beyond recognition. No aspect of society, economy, and politics of rural areas has remained unimpressed by this phenomenon.

Despite the availability of a vast reservoir of studies on rural transformation, social relations as a contributing factor have been conspicuously absent. Not many studies on transformation have underlined that outmigration from villages and withdrawal from agriculture might be triggered by the Dalits' desire to escape caste domination of landed upper *jatis*. Women's urge to liberate themselves from the controls of rural families has not found much attention in social science research on migration. Social structure might not have figured prominently in social science research on transformation, but its close interface with the latter needs a thorough probe. It entails locating the dynamic of transformation at the intersection of social relations, the market, and the state.

Methodology

The Giri Institute of Development Studies, Lucknow, initiated a series of village surveys in 2012–2013. The first household survey to collect data from four selected villages was conducted in 2013. The baseline survey laid the basis for future studies. The second survey of this study was conducted in 2020, seven years after the first survey.

The First UP Rural Transformation Survey (2013)

The baseline survey was conducted in the summer of 2013 in four selected villages in four different regions of the state. These villages include Seemli (Muzaffarnagar, Western), Gohanakallan (Lucknow, Central), Pandari (Chitrakoot, Bundelkhand), and Senapur (Jaunpur, Eastern).

The 2013 survey was the first in this continuous series of studies. Since it was only one-point data, it could only provide baseline data and status reports. But while selecting sample villages, those villages that were studied by social scientists earlier were chosen. The existing literature on these villages provided valuable background information. The availability of previous studies also allowed comparison of those investigations with current studies on some aspects qualitatively, if not quantitatively.

This exercise yielded insights into changing land and labour relations, penetration of technology and productivity in agriculture, agrarian distress, indebtedness, diversification of employment, migration, financial inclusion, caste relations, gender relations, education, health, and the impact of government schemes like MNREGA.

The First Rural Transformation Survey collected data on six significant aspects of rural life. Demographic data investigated religious and social group-wise population, sex ratio, age profile, educational and occupational status of the people, etc. Ownership of assets included livestock assets, agricultural implements, land, buildings, machines and pieces of equipment, durable household assets, and financial assets. Data on agrarian change mainly examined social group-wise landholdings, land leasing practices, land use patterns, sources of irrigation, the composition of input, cropping patterns, and agricultural output and disposal of produce.

The fourth aspect extensively covered by the survey related to household income, expenditure, indebtedness, and living status. This section attempted to estimate income from different sources. Data on debt was collected on several indicators like sources, period, purpose, and outstanding amounts. Living standards have been assessed based on housing, electricity, water and sanitation, and cooking fuel access. Data on employment conditions of agricultural and non-agricultural workforce in different employment sectors, self-employed workers, salaried workers, migrant workers, access to social security, etc. enabled analysis of employment and migration status of the rural population. Data on indicators like access to education, health services, Public Distribution System (PDS), and political participation facilitated analysis of the social development component of the First Rural Transformation Survey.

Four villages reflected different patterns of transformation. The landownership pattern of all four villages was marked by decreasing the size of land holdings while maintaining inequalities between social groups. Land leasing practices varied from village to village. At the same time, chemical fertilisers, pesticides, high-yield variety seeds, irrigation, and labour-displacing equipment were on the rise. Yield levels of wheat and rice were also comparable to Indian averages. Though average livestock asset ownership was highest for "general category" households, followed by OBCs and SCs, a larger share of income for OBC and SC households came from animal husbandry rather than from agriculture. This might relate to the relatively lower income of OBC and SC households from agriculture.

Diversification of employment to non-agricultural sources has been a frequently discussed phenomenon in social sciences. These villages were no exception to this trend. The First UP Rural Transformation Survey identified three significant avenues of employment and income sources where this shift was taking place: construction, self-employment, and government scheme work. Classification of workers into static categories, such as rural/urban, agriculture/construction, migrant/local, etc., was becoming increasingly difficult for social scientists. The rotation of workers between different jobs at different places in a year had become such a prevailing phenomenon that it seems to have converted a larger part of workers into a "floating population". Wages in the non-agricultural sector were higher, but much less than minimum wages and workers' bargaining power related to household landownership.

Between this quest, an improvement in social development indices, including education, health, asset ownership, and housing of all social groups, was reported, but inequality persisted. The analysis also documented that income-based inequalities were far greater than consumption-based inequalities. The persistence of these inequalities along the social hierarchy reflected a lacuna in the design of the development strategy adopted by the Indian state after independence.

The Second Uttar Pradesh Rural Transformation Survey (2020)

This longitudinal study examined agrarian changes in landownership, tenancy relations, cropping patterns, productivity, and technology penetration. It also investigated the dynamics of employment diversification and migration and explored the pattern of household asset ownership, income, indebtedness, and living status. It also studied socio-political transformation, including caste relations, political participation, and gender relations, and probed progress on social development issues like demographic changes, access to education, health, food nutrition, housing, and the impact of government schemes.

Selected Villages and Sample Households

The First UP Rural Transformation Survey covered all 1638 households in four villages in four distinct regions of the state. It included 503 households in Gohanakallan (Lucknow), 471 households in Senapur (Jaunpur), 368 households in Pandari (Chitrakoot), and 296 households in Seemli (Muzaffarnagar). In seven years since 2013, the number of households has increased due to some households splitting. All these original and newly constituted households were covered in the second survey.

Data have been collected on demographic aspects, ownership of assets, agrarian change, household income, expenditure, indebtedness, living status, employment, migration, and social development.

Similarly, data on social group-wise landholdings, land transactions, land leasing contracts, land use patterns, sources of irrigation, cropping patterns, and agricultural input–output and disposal of produce have been gathered from the household survey.

A study of employment diversification entailed data collection on various indicators such as wage employment availability and agricultural labour's daily wages. Similarly, data on the employment conditions of the non-agricultural workforce in different employment sectors, wages, self-employed workers, salaried workers, and access to social security have been collected. The household survey also gathered data on the composition of migrant workers, duration of migration, destination of migration, etc., which would help analyse migration patterns. To estimate "Ownership of assets", the survey gathered data on livestock assets, agricultural implements, land, buildings, machines and pieces of equipment, durable household assets, and financial assets.

The socio-political transformation could be understood by analysing both quantitative and qualitative data. Quantitative data could reveal inequality along caste and gender lines. Qualitative data could explore power relations based on caste, gender, class, etc. Demographic aspects could be examined by looking at the educational and occupational status of the population, besides usual indicators like social group-wise population, sex ratio, age profile, etc. Data on indicators, like access to education, health services, PDS, political participation, the impact of government schemes, etc., have helped analyse state intervention for people's welfare. In addition to the household survey, qualitative data have been collected through detailed field notes.

The work has been done in a comparative framework (Moore Jr, 1967). An analysis of data collected in 2013 and the 2020 survey revealed interesting aspects of transformation in rural areas during these seven years. Besides, longitudinal changes, the study also attempts comparision between villages to understand how individual villages have been changing differently given their distinct history, political economy and sociology. This comparison has been made between household data aggregated at the village level. The data collected from the household survey has been analysed along the axes of class, caste, and gender. Since these villages fall in different regions of the state, they also divulged regional differentiation.

Reference Period

The Second Rural Transformation Survey (2020) was conducted from October 2020 to December 2020. The survey data was considered a cut-off mark for household data on demography, education, employment, asset, land, etc. The agricultural year 2019–2020 was taken as the reference period for agricultural operations like crop production.

However, the seven-year period might be considered relatively short for a longitudinal study, but the data shows that this period has witnessed unprecedented churning in occupational patterns, demography, educational scenarios, and inequalities among social groups. Social science research could not afford to miss the opportunity to study the dynamics of social change engulfing the countryside.

Sample Households Surveyed in 2013 and 2020

Due to this peculiarity of the data set, only those households surveyed in 2013 and 2020 have been considered for the analysis in the chapters. This goes without saying that a newly created household due to the split has been included in the analysis if the original household was already on the panel of comparable households. The split households were treated as part of the original households in data analysis. Finally, we have a panel of 1348 households surveyed at two points.

Senapur, the eastern UP village, appears to be a Dalit preponderant village. Around 55 per cent of the households of a panel of 356 comparable households belonged to the SC category in this Jaunpur village (Table 1.1). Another 31 per cent reported OBC as their social group, while around 14 per

Table 1.1 Social Group-Wise Distribution of Sample Households Surveyed both in 2013 and 2020

Social Group	*Village*				
	Senapur	*Seemli*	*Pandari*	*Gohanakallan*	*Total*
SC	54.78	41.41	43.85	37.01	44.07
ST	0.28	0.39	0.33	0.23	0.3
OBC	30.62	57.42	42.86	51.03	45.03
Others	14.33	0.78	12.96	11.72	10.61
Total	100 (356)	100 (256)	100 (301)	100 (435)	100 (1348)

Source: Primary Survey (2013, 2020).

cent were from the non-SC–OBC–ST category. Unlike Senapur, in the village under study from the Muzaffarnagar district, OBC emerged as the largest group in Seemli, with over 57 per cent of households in a panel of 256. In contrast, 41 per cent of households belonged to the SC category. The presence of other social groups was negligible in this western UP village.

On a panel of 301 households, both SC and OBC social groups had an almost equal presence in Pandari village of Bundelkhand. Around 44 per cent of households in this village of Chitrakoot belonged to the SC category, followed by 43 per cent of OBC households and 13 per cent of non-SC–ST–OBC households. In the selected village from central UP, Gohanakalla's social group-wise composition was slightly different, with 51 per cent OBC households, 37 per cent SC households, and 12 per cent non-SC–ST–OBC households on a panel of 435 households. In all these villages, the presence of ST households was negligible.

Among the four selected villages, Pandari and Gohanakallan were inhabited by Hindus only (Table 1.2). All the households in these villages reported their religion as Hinduism. In the other two villages, a relatively smaller number of households followed Islam. In Senapur, around 6 per cent of

Table 1.2 Religious Group-wise Distribution of Sample Households Surveyed both in 2013 and 2020

Villages/Religion	*Hindu*	*Muslim*	*All*
Senapur	94.1	5.9	100
Seemli	79.3	20.7	100
Pandaari	100	0	100
Gohanakala	100	0	100
Total	94.51 (1274)	5.49 (74)	100 (1348)

Source: Primary Survey (2013, 2020).

households reported themselves as Muslim, while over 20 per cent of households in Pandari reportedly followed Islam. In all four villages, nearly 95 per cent of sample households surveyed in 2013 and 2020 followed Hinduism, while the remaining 5 per cent followed Islam.

About the Book

As the Global South undergoes overwhelming transformations, departing from the path prescribed by mainstream social science theory, the work adopts an alternative view of the complexities involved in the interface of caste and gender with demographic changes. It investigates occupational diversification and rising inequalities amidst closer integration between the rural and urban, presenting new challenges and opportunities. Based on longitudinal data from four villages in the northern Indian state of Uttar Pradesh, including Bernard Cohn's Senapur and D N Majumdar's Gohanakallan, the study indicates that demographic shifts offer an opportunity, but social development deficit emerges as the major challenge. The work debunks the prosperity theory of the withdrawal of women from the workforce, highlighting the nuanced dynamics of women's displacement. The work also reflects the increasing separation between the site of production and reproduction amidst rural-urban coalescence. The study unearthed a high degree of landlessness among Dalits and backward Muslims and women's continued exclusion from ownership claims over land despite increasing participation in and contribution to the agrarian economy. The study also explored the shifts in the agricultural economy, productivity, technology deployment, and social dynamics. While land-caste congruence withstands headwinds of change, the resulting imbalance is evident in rising land tenancy. The work culminates in identifying policy gaps and recommends steps for a sustainable and inclusive rural transformation in the Global South.

After Chapter One engages with the debate on rural transformation and identifies the gaps in the framework within which discourse has been conducted, Chapter 2 describes the selected villages: two extensively studied in the 1950s and two new villages. Bernard S Cohn's Senapur and the legendary anthropologist D N Majumdar's Gohanakallan offer an in-depth exploration of the 'functional relationship' among the castes, specifically caste and family, culture, economy, legal systems, politics, and many other facets of village life. Alongside the research sites chosen by Chicago and Lucknow schools, the present study interrogates artificial boundaries between social relationships and production relations.

Based on longitudinal data, Chapter 3 iterates that demographic changes have driven the transformation of rural areas, and access to quality education has shaped the dynamics of change, contributing to inequality among social groups. The rising asset inequality in rural areas was only the tip of the iceberg as a larger share of professionals settled down in towns than manual workers who formed the bottom of the hierarchy in both locations.

Chapter 4 identifies complexities beyond the conventional binary of agricultural and non-agricultural occupations. The assumption of a predetermined and linear progression has been abandoned in favour of a more nuanced approach, recognising the multiplicity and simultaneity of occupations.

Chapter 5 notes that the landownership structure has proved to be the most resistant to change, and caste-gender continues to have a grip on land relations. The analysis showed that the contemporary 'small tenancy', though on the rise, is not necessarily between a landless or marginal peasant and a large owner but often involves 'parallel' lease contracts between similarly placed landowners. The dormant land sale market does not typically imply that land has yet to be commodified as social and political relations closely interface with land ownership in rural areas, but these equations increasingly materialised via market and democratic set-up.

Chapter 6 on agriculture is concerned with changing cropping patterns, productivity, and production and whether agriculture was moving from primarily a subsistence activity to a market-oriented enterprise. The most crucial transformation noted by this work concurs with recent studies on the disjunction of 'rural' and 'agriculture'.

Chapter 7 Seven years after the first survey, we have seen further expansion of the tendency to integrate the rural economy with the market economy. However, the most interesting aspect of the transformation rural areas experienced was the modification of pre-existing relations rather than a complete replacement. The data collected by both surveys revealed that workers rarely quit the agriculture sector but diversified outside it while maintaining a toehold in it. Villages might have lost their status as sites of production and accumulation but continued to play the role of reproduction.The capital-caste complex provided opportunities for horizontal movement and restricted vertical movement. Gender emerged as a strong determinant in occupational diversification as female workers appeared to fill the space vacated by male workers while migrating to newer jobs. The demographic transition and transformation of the rural areas are linked via their association with capital. Rising inequality was another remarkable feature of the transformation of rural areas under the impact of market integration.

Based on a comprehensive exploration of numerous facets of rural life, the study identified three sites for policy intervention on a priority basis. First, the realisation of the 'demographic dividend' couldn't be left to the vagaries of the market, and urgent state intervention in education, health, and social security was needed. A related site for attention would be occupational diversification, which has wholly bypassed women. A similar enabling role of the state is needed to neutralise caste-based handicaps to facilitate vertical mobility rather than horizontal movement, as recorded by this study. It has been recognised that the agricultural economy and, more importantly, people dependent on it are seeking a way out of the crisis. The two pain points-credit and marketing- needs to be addressed. The most challenging task before public policy in the Global South would be to restore rural areas' productive capacities. This will bring employment opportunities closer to workers, especially women. The policy must emphasise that the non-farm sector could go beyond food-processing industries. It could produce anything conceivable on earth, not necessarily in large capitalist industries.

2 The Transforming Rural Landscape

The Story of Four Villages

The Village as Research Site

Whether the "village" constituted a "sociological reality" has rattled Indian sociology. Metcalfe's notion of the "little republic" and Dumont's argument of Indian society as a whole being the true object of sociological enquiry provided vantage points to this debate. Dumont and Pocock (1957) denied any significance to the village as a research site since they saw India as one civilisational entity, villages having no meaningful place in it. Their challenge was particularly directed against the flourishing village studies tradition engaged in juxtaposing the "field view" against the "book view" in the newly independent country. On the other hand, Bailey (1959) was interested in distinguishing between Sanskritic values and the everyday life of the people inhabiting India's diverse landscape (Sharma, 1969).

Both the Metcalfe and Dumont approaches carried an essentialist streak; the first took India's villages as static, self-contained, and self-sufficient republics, and the second considered the influence of Sanskritic values as binding on people's social life. The Western gaze was unmistakably present in both approaches. The forceful imposition of the idea of an "isolated whole" on an Indian village essentially emanated from the anthropological explorations of primitive groups. The second group of sociologists came looking for villages that had completely lost their autonomy in the fashion of the coalescence of European and American agricultural economies with a broader capitalist economy. And more than that, they resisted change in their own ways. In the first view, the villages have been an unchanging reality since time immemorial, surviving massive political upheavals without losing their essence. And the second one saw the unabated persistence of civilisational values. While looking for a peasant village that constituted a "whole" on the one hand, and a village that had completely lost its autonomy on the other hand, the social anthropologists were reflecting on the limitations of their discipline rather than the question of the village as a research site (Sharma, 1969; Jodhka and Simpson, 2019).

The main question that came in the way of recognising the village as a research site was whether it had a "sociological reality". Dumont and Pocock (1957, pp. 10–18) objected to the practice of village studies on the grounds

DOI: 10.4324/9781003588245-2

that (1) the territorial factor was only second to kinship and caste in social organisation, (2) India was not composed of villages in the sociological sense, (3) villages in India lacked solidarity, (4) villages could not be conceptualised as a whole in itself, and (5) Indian society unified by civilisational values was the true object of study for sociologists.

Dumont and Pocock received a massive response from social anthropologists practising the single village research site method. Fortunately, Senapur, a selected village for the present study, was a major site of these contestations. While blasting the notion of the isolated village by describing its connection with the outside world in the form of economic transactions, kinship ties, landownership, religion and social reform movements, and migration for jobs and education, Opler (1956, p. 10) did not forget to underline that "extensions" of a village should not be misconstrued as the absence of a separate identity and cohesiveness of the community.

Responding to the question of whether a village was characteristically a functional and distinguishable social unit, another study on the same village, while not discounting intra-village factionalism and inter-caste struggles, underlined the village deity, legal recognition of the village as a territorial unit, panchayat (local self-government), *jajmani* and caste relations, and the agricultural land tenancy system as sites that bound the village together (Singh, 1956).

Historical evidence from the region also emphasised that the village was the basic socio-economic unit of a larger socio-political region, often referred to as *pargana*, along with the intermediate layer of *tappas* (Singh, 1968). The cluster of villages formed what Cohn termed a "little kingdom", the basic jural unit. In the case of Senapur, the region was called *dobhi taluka*, consisting of about 100 villages spread over a 40-square-mile area. The point Cohn was making to address the problems that arose when anthropologists moved from the study of primitive communities to the social units of the great civilisations was that both the extreme views on the village's autonomy were ahistorical. The village certainly enjoyed self-sufficiency in some aspects, but the notion of autonomy was not static but articulated in constant negotiation with the outer world (Cohn, 1959).

The Gohanakallan narrative of Majumdar et al. (1955) also debunked the notion of isolated villages while providing rich data on the inter-caste relations within the village. This is the second village selected for the present study. The Majumdar village appeared knitted by *jajmani* relations, and the system ran on the exercise of power by the dominant landowning caste. Most of the studies of the period largely focused on cohesiveness and functional relationships between the groups constituting the village and hardly paid attention to the role of power in holding various components together.

The distortion in the perspectives on the village also emanated from the persistence of seeing it only as a residential community. Rather, it should be seen as an enterprise akin to modern industrial townships. Social scientists observed villages as existing rather than becoming. As elsewhere, many villages in eastern Uttar Pradesh have been established rather than existed. The 17th and 18th centuries witnessed new lands being broken and villages established by entrepreneurial landlords. The process of "colonial construction" might have gained momentum with the escalating pressure on land. In addition, *zamindars* in eastern Uttar Pradesh offered concessions to the peasantry to settle them in their area, seeking enhanced agricultural production and optimal utilisation of land resources. These concessions might include free land for house sites, payment of revenue arrears in interest-free instalments, and assistance during natural calamities (Rizvi, 1988). There were other ways of establishing new villages. For instance, Majumdar et al. (1955) noted two satellite villages around Goahanakallan coming into existence after 1862.

These enterprise-looking villages would enjoy self-sufficiency in some aspects and extensive connections extending to "little republics" and beyond. The measure of autonomy and dependence would be contextual, often contingent upon human and natural resources. The foremost feature of self-reliance that any member of a village community would flaunt was the number of inhabiting castes. The higher number of castes signified the availability of a larger number of specialised trades in the village. For instance, Senapur had 23 resident castes against 15 of Gohanakallan. However, the latter would not be considered in bad condition, but the former was surely a more resourceful village. Any village missing a particular trade would eventually depend on neighbouring villages. Often, older villages possessed more diversified skill sets.

Land formed another crucial measure of self-sufficiency. A prosperous village would often possess more land than it could cultivate. Conversely, a land-deficient village would depend upon others for leasing-in agricultural land. Mostly in a cluster of villages, one particular village was considered the main village where large landowners resided. This village owned a larger quantity of land than the villages where peasants, service, or labouring castes resided in large numbers. Occasionally, the legacy owners residing in one village owned land in other villages. A shortage of labourers would equally affect the economy of a village. They would have to depend on the supply of labourers from neighbouring and, in some cases, from far-off places. The land and labour-surplus villages frequently formed an enduring exchange relationship, though not on an equal footing.

However, this is not a place to spell out connections between a village and other places; earlier works have provided vivid details. Often studies customarily cite the "self-sufficient economy" of the old days, but the research-based arguments recognise that almost every village found a place in the sprawling trade networks that spread across the country and beyond. Senapur was no exception, where traders from distant places like Punjab came to sell commodities. During the colonial period, industrial goods started selling in village markets, beating local handicrafts (Opler, 1956).

The perpetual fluidity between autonomy and "extensions" of a village demystified basic terms of the debate on sociological reality. Contrary to the hopes of Dumont and Pocock, research has shown that villages had a crucial role in influencing the social life of residents. Whether villages constituted a "whole" became immaterial in the face of evidence on the village as a site of social relations, the everlasting interest of sociology. The question of "solidarity" was only a legacy of anthropology's long association with primitive tribe studies. Peasant villages were marked by conflict and solidarity. Anybody looking for a homogenous community in these villages would be greatly disappointed (Srinivas, 1951; Lewis and Barnouw, 1956; Bailey, 1959; Marriott, 1955; Mayer, 1960). A short description of the studied village would further clarify this point.

The critiques of Dumont and Pocock have not been able to breach the essentialist framework that saw the village as a homogenous entity. Lewis (1958, p. 318) fittingly pointed out that "people didn't just live in villages; they belonged to it". In earlier works, too, the village found a central place in the kinship network and a "focal point of reference for individual prestige and identification" (Marriott, 1955, 178) and "a sense of ritual and social integrity, and political solidarity against the outside world" (Srinivas, 1951, pp. 1051–55). However, these empirical observations might not necessarily be disputed, but the village's treatment as a "whole" unit would strike first an onlooker. There was hardly any effort on the part of these classic works to probe the perception of the village among its segments. The de-homogenisation of the village by exploring the differential experiences of the "periphery" and the "core" would have produced a richer sociology. For instance, whether Dalits had a different take on the "integrity" of the village than the dominant caste? Or did women identify with the village in the same way as others?

Beyond the rather unproductive debate on sociological reality, anthropologists found villages useful as "observation-centres" where macro processes could be closely observed. One could also study the interaction of social, economic, and political forces (Srinivas, 1955; Beteille, 1996). There was a common assumption that sociologists and anthropologists considered villages themselves as an object of research, while economists took them as a site of research, representing rural areas (Himanshu et al., 2016, p. 2). This was a case of mistaken identity by collapsing a variety of sociological traditions together and the evolving nature of the discipline. In fact, sociologists were attuned to the social relations within the village and the way they extended beyond it (Chakravarti, 2001). For instance, caste equations were often seen as a product of a combination of *jatis* residing in a particular village, mutually engaged in social, economic, cultural, and political spheres, particularly when seen alongside the political economy, but caste networks extended far beyond the village boundaries. Gradually, the discipline overcame its initial tendency to view the village as a watertight compartment and shifted focus on "extension" while not losing sight of village-specific relations, which enabled them to study the village, both as a research site and microcosm of rural life and connections between both the levels.

This understanding of villages as a research site gave much-needed impetus to subsequent studies in various disciplines, including economics, political economy, sociology, and political science. The methodological shift, which viewed villages as microcosms of predominantly rural southern societies, enabled social scientists to achieve a similar level of precision in data collection and generalisation that was only available to natural scientists in lab settings.

Notwithstanding their limitations, given diversity and externalities, these exercises have proven useful to various disciplines in generating new hypotheses and validating existing theories. This methodological development also proved useful for policy-oriented development studies, where data from a small set of villages came in handy in identifying rural problems and areas requiring state interventions. Once a methodological impasse concerning the village as a research site was overcome, a rich village and rural studies tradition probing numerous villages across India thus emerged. The two villages chosen for this study, Senapur and Gohanakallan, were among them.

Senapur

One of the most researched villages in north India, Senapur came to the academic limelight due to the efforts of Bernard S Cohn, Morris E Opler, Rudra Datt Singh, and other anthropologists. Their interest in village studies was so great that Cornell University established a field station in eastern UP. They studied social relations, specifically caste and family, culture, economy, legal changes, politics, and other aspects of village life.

Senapur was a large village, having a population of 1852 and an area of 1047 acres in the early 1950s. Twenty-three castes inhabited the village, but Thakurs, Chamars, Ahirs, and Noniyas were preponderant. The numerically largest caste of Chamars had 636 persons, followed by 436 Thakurs, 239 Noniyas, 116 Ahirs, and 67 Lohars. Other castes were numerically insignificant, having less than 50 persons each. Half of the castes residing in the village had less than 20 persons each. The Senapur studies argued that Indian village life might appear structured from the outside but was in constant flux. They further argued that, against popularly held beliefs about the caste system and related behaviour patterns, the social organisation in rural India often responded to external stimuli (Opler and Singh, 1952).

Opler and Singh's work demystified the popular notion of Brahmin supremacy in the caste hierarchy. As in Majumdar's work cited later, they, too, point out that being economically dependent on others, Brahmins, at best, receive ritual regard. They were not influential in any village matter. The Thakurs, the biggest landowners, wielded real power in the village. They directly controlled 82 per cent of the village land and owned a substantial portion of the remaining land in other forms. The village had a huge population of families dependent on Thakurs for leasing land. The remaining castes, mostly occupational and labouring households, sought a livelihood by providing services to the peasantry and Thakurs (Opler and Singh, 1952).

In terms of numbers and assertive attitude, Ahirs stood out from the rest of the population. They revealed the signs of restlessness against the domination of Thakurs. And they were the most organised group in the village. On the other side, Noniyas offered passive resistance in the form of non-cooperation. But in both the cases and the case of other castes, whenever there was an opposition, one party was always the Thakurs. The power tussle in the village was confined to these three groups, and Chamars, whenever they asserted, were mainly about their wages and better treatment from the employer. Their way of pressing for their demands was also to resort to boycotting work. All other groups, mostly occupational castes in *jajmani* relations, consciously abstained from village politics (Opler and Singh, 1952).

All of this started changing with India's independence in 1947. Opler and Singh (1952) observed that everybody else, except for Thakurs, was excited upon the realisation that they could influence the formation and functioning of the village government. They also underlined that the poor were particularly hopeful after hearing that they, too, had the equal right to vote as the wealthy people of the village. Opler and Singh further observed that all this culminated in the village government elections of 1949 when lower caste people of the village collectively posed a challenge to Thakurs under the leadership of Ahirs and Noniyas and with the support of some Thakurs of low economic status, and Brahmins defeated the powerful in the election. The story of Senapur had taken a turn with the country's history, reemphasising the futility of the notion of village isolation.

Opler and Singh (1952) found caste status, numerical strength, and power rooted in landownership. It was not an exception, but Senapur presented a picture-perfect model of an upper caste controlling almost the entire land and rural power structure. However, it did not deter the village's farmers from engaging in intensive cultivation. The authors further pointed out that digging several wells for irrigation ensured they could grow several crops, including sugarcane.

It was not a new development. The process was set in motion in the 19th century after the advent of British rule and its policy of strengthening the position of zamindars as the social base of colonial administration in rural areas. Thakurs also utilised the opportunity thrown up by newly introduced cash crops of indigo and sugarcane. The new stream of income further cemented their traditional hold over rural society. In the 19th and 20th centuries, the rising population and destruction of the artisanal economy increased pressure on land, coinciding with inflation in the prices of agricultural commodities, turning land into an asset. By this time, Senapur was already drawn into a global commodity trade circuit that, on the one hand, dumped industrially manufactured goods from Europe, replacing locally produced goods, and, on the other hand connected village agriculture with international trade, along with its pitfalls (Cohn, 1959).

Studying Senapur, Cohn (1959) "seriously doubted" the idea of isolation and the static nature of villages since famines and warfare would have occasionally altered their social structure. And the pace of change quickened during the colonial period with the opening of job opportunities in trade, government services, and newly established industries, catalysing economic and social change through closer contact with the outside world. Cohn added that the established patterns of relationships received a major jolt after independence, with the new political, economic, and legal framework taking shape.

Cohn (1959, p. 89) argued:

> Land reform has altered the relationship among Thakurs and their dependents, and the nationalist struggle, democratic elections, and movement for social and economic uplift of low castes have destroyed the moral base of the relationships of superordination and subordination among the Thakurs and their low caste dependents.

The stream of studies on Senapur minutely documented the village's connections with the outer world, debunking any proposition about its isolation. Against the popular view about Indian villages, Opler (1956) maintained that villages could still be economically isolable and self-sufficient, but socio-cultural connections have been more pervasive. Opler further demolished the preconceived notions about the village's outer reach with the advent of modern developments. He pointed out that Senapur's relations with far-off communities were based on traditional practices, common ancestral origin, and village exogamy. None of these were new things in the social life of Senapur residents.

One of the interesting differences between social and economic outreach, marked by Opler (1956), was that economic agents often replaced the previous ones. For instance, Opler noted that the village's integration with the world commodity market via local marketplaces and the arrival of industrial goods happened at the expense of vendors from distant places, often from other states. Opler (1956, pp. 9–10) remarkably said that "often it is the pattern of outside contact rather than the fact of outside contact which is altered...the relations with the outside are retained but the mechanism which helps maintain their change in character". Despite these connections, Opler emphasised, the village was not disintegrating. Most of the workers who migrated to urban areas for work did not plan to settle down there.

Another substantial change that reached the village with India's independence after a long struggle against colonial rule was the decentralisation of power as a declared state policy seeking to strengthen village-level decision-making on matters affecting its residents. It included local courts to settle disputes in a circle of villages, democratically elected governing bodies, and managing committees. The recognition of the village as the unit for policy intervention and many other legal and political initiatives of the state and union governments further cemented its unity (Opler, 1956).

One more notable feature of Senapur was its longstanding connections with the outside world for employment, going back to the middle of the 19th century when village inhabitants of the Chamar caste worked in British households. In the 1950s, when American anthropologists studied the village, 36 members of the Dalit community were working outside the village, mostly in the jute mills of Calcutta (now Kolkata), the cotton mills of Kanpur, and the coal fields of West Bengal, and as industrial workers, tonga drivers, and rickshaw pullers in Delhi, Bombay (now Mumbai), Banaras (Varanasi), Cuttack, etc. Besides these menial workers, some were also engaged in white-collar employment. Most adult men among Dalits had worked outside the village without reconsidering the desire to return when some cash required to meet urgency at home was earned (Cohn, 1958).

The Senapur studies proved to be a graveyard for the theories of modernisation and westernisation. An example from Cohn (1958, p. 416) illustrates this:

> It is usually thought by observers of Indian life that urbanization and industrialization will be main contributors to the Westernization of Indian life; however, there are a few indications that urban employment might be an influence in the Camars' [Chamar] taking over a more traditional Hindu ritual pattern.

In support of the argument, Cohn (1958) observed that Dalits did not have much exposure to Hindu traditions in the village because they were seen as untouchable and sanctioned against hearing Vedas. Village Brahmins did not preside over their rituals. More importantly, Cohn underlined that it was only after India's independence and the country adopting the Constitution of India that Dalits were permitted entry into temples. The imposition of these conditions kept them away from the "Great Tradition". In urban settings, Cohn argued, these people engaged in several activities, including building their own temples and singing *bhajans* (religious songs), consequently coming closer to orthodox Hindu traditions.

Modern education proved to be another instrument for bringing Dalits closer to "The Great Indian Tradition". First, after Independence, when Dalit children started attending the village school, the newfound literacy enabled them to learn about Hindu traditions, something they knew relatively nothing about previously, except for participating in some festivals and watching the yearly enactment of "Ram Leela". The newly educated Dalits also started to look down upon the traditions of Chamars for deviating from the orthodox Hindu traditions (Cohn, 1958).

However, the impact of modern education on Thakurs was very different. They received higher education and were exposed to Western civilisation for being adept in English. They joined modern occupations like college teaching, state bureaucracy, and Western-style businesses and relocated to modern cities. While in the case of Chamars, access to modern education strengthened orthodoxy among them, it catalysed Westernisation among Thakurs (Cohn, 1958).

One more theoretical myth that Bernard Cohn diffused by drawing data from Senapur was about modernity and the past. Cohn (1961) cited Edward Shils, who maintained that traditional society could be identified based on a strong attachment to the past that determined behaviour patterns. Disputing Shils's claim, Cohn contended that modern society would not be bereft of the past, but it would be largely common among caste communities, and the shared past could determine and validate behaviour patterns.

Seven decades down the line, not everything has changed in Senapur. The Thakurs of the village continue to own large tracts of agricultural land, so much so that they cannot cultivate independently. Due to their very high educational achievements, many Thakur families migrated to far-flung places, including Mumbai, Delhi, Lucknow, Gujarat, and the northeast.

Being an important centre of India's First Freedom Struggle in 1857, the village's fame spread far and wide in the district and adjoining areas. There was a *shahid smarak* (martyrs' memorial) in the village in memory of those who sacrificed their lives for the country's independence. Most freedom fighters whose names were recorded in history belonged to the Thakur and Yadav castes. Chamars of the village felt aggrieved for omitting names belonging to their caste. With slight modifications of administrative boundaries since Cohn's studies, Senapur village was in the block and sub-district Kerakat of Jaunpur district in eastern Uttar Pradesh, almost 40 kilometres from the district headquarters.

A large part of Senapur's agriculture was a low-lying area where waterlogging was a significant problem. After a good spell of rain, all these lands and a part of the habitation got submerged underwater, and the absence of any drainage system ensured that this condition prevailed for a long time, damaging standing crops. Farmers started transplanting paddy saplings before the start of the rain to allow them to grow taller, choosing a variety of paddy that grew taller and transplanting it only after it got taller. Even after these efforts, farmers suffered huge losses due to waterlogging.

The mechanisation of agricultural operations was taking place at a pace in the village. A harvester was often deployed to harvest crops, whereas a seed drill became the preferred choice for seeding. One farmer in the village also owned a harvester-combine, which was hired in and around the village. The village was electrified very early in the 1970s, but the electricity supply was erratic during the survey in 2020.

During the Kharif season, paddy, corn, millet, *arhar* pulses (tur), urad pulses, etc. crops were cultivated in Senapur. Wheat, gram, matar, mustard, potato, onion, etc. were major Rabi crops. Some farmers also cropped *chari* for animal feed. Sugarcane was a minor crop used only for making *gud (jaggery)*. A minor canal passed through the village, but, leaving plots near the canal, private tube wells and diesel pumping sets formed the major source of irrigation for all other fields. This enhanced the cost of cultivation.

About two-thirds of the village population belonged to the SC category, followed by OBCs constituting one-fifth, and the remaining one-sixth from the non-SC–OBC–ST category. Some Muslim families belonging to the Dafali caste also lived in the village. This *jati* comes under OBC. Among SCs, Chamars were the largest group, whereas Yadav, Lohar, Badhai, Kumhar, Kahar, Teli, and Lonia constituted OBC. Thakurs formed the non-SC–OBC–ST category in the village.

Unequal land distribution was also reflected in widespread landlessness among Chamars and tiny plots of land among OBCs. This imbalance in the ownership of agricultural land in the hands of one group and cultivation as a source of livelihood for another resulted in widespread land tenancy. The land-poor groups leased land from the landed classes and hired themselves as manual labourers for survival. Wages for women agricultural labour generally fluctuated in the band of Rs 150 to Rs 200, while men labourers fetched between Rs 250 and Rs 300. Masons earned Rs 500 on construction sites, whereas headload workers' daily wage was around Rs 300 in 2020.

There were four major prevailing systems of tenancy in the village. First, both parties shared inputs and outputs (main and by-product) equally, except labour provided solely by the tenant. Second, the entire cost of cultivation except labour was provided by the landowners, who took two-thirds of the main products and all of the by-products. The tenant kept one-third of the main products for her labour. Third, except for tilling and labour, both parties equally shared the remaining expenses, and the primary produce was distributed equally; the tenant kept all the by-products. The fourth tenancy system was cash rent, under which the lessee paid a fixed rent.

Besides small tenancy, other members of the households migrate to other states and cities to work as manual labour because of the local economy's scarcity of jobs and low wage rates. Landless families of the village had no other option but to work as manual labour. Women in the village worked as agricultural labour, whereas men migrated to other states. A prior acquaintance from the village and nearby area mostly played a role in the migration network.

In some years, OBCs and SCs had also obtained government jobs such as police, army, teachers, and some officers. This change was also reflected in their increased income and living standards.

This village had an intermediate college for a long time alongside a primary school, an upper primary school, and a private school. This way, facilities to obtain education up to the 12th standard were available in this village. The availability of educational infrastructure was also reflected in the heightened awareness about education among all social groups in the village. Many village elders were also educated, and children's education received adequate attention. At the time of the survey, many children were enroled in courses like B Ed, BTC, B Tech, etc.

Girl children were also provided education. Many families were sparing girls from household chores to attend school and work outside the village.

Girls also felt confident and free in their movement outside the village. Child marriages were at an all-time low, if not entirely abolished from the village.

Caste discrimination against Dalits was still rampant in Senapur. If OBCs or non-SC–ST–OBC families invited Dalits to their feasts, they would be served food separately. And in the feasts of Dalits, people from other social groups would avoid eating. Dalits generally invited only those people from non-SC–ST–OBC households whose land they were cultivating on lease. While most Dalit families were under the domination of Thakurs owing to their status as insecure tenants, some families were essentially under the control of the erstwhile landlords. Besides tilling their land, women from these Dalit families did household chores and cared for animals in Thakur mansions. Alcoholism was a significant issue impacting the village, especially among Dalits. Women felt harassed and agitated by this issue.

Government schemes had a limited impact on the village. Almost half of the population of Senapur was still open defecating despite the construction of toilets under Swacch Bharat Abhiyan. Several families still lived in dilapidated houses due to not receiving benefits under Pradhan Mantri Awas Yojana. People were also facing much corruption in PDS. The Health Insurance Scheme also had limited access in the village.

Gohanakallan

The village was extensively studied by legendary anthropologist D N Majumdar and his colleagues at the University of Lucknow in the 1950s. They provided a detailed description of the "functional relationship" among the castes of the village. They found the predominantly agrarian village organised by "symbiotic arrangements" between castes divided into watertight compartments based on social status. Each caste was subjected to certain norms regulating their behaviour with other castes, and these rules were upheld by the control of the "upper caste", in this case, the landowning Thakurs, over the lower caste. Majumdar et al. (1955) believed that the "caste system" had a more profound grip over rural society than urban society. The arrangement, marked by upper caste leadership and economic interdependence among castes, kept conflicts under check and sustained the cooperation necessary for a self-sufficient politico-economic unit like a village.

Based on data collected on the numerical strength of castes, Majumdar et al. found that Chamars were the largest community in the village, followed by Rajputs, Pasis, and Ahirs, and 11 other minor groups, mostly service castes engaged in carpentry, iron-smithing, washing clothes, rearing animals, hairdressing, religious services, etc. Thakurs were the most powerful group, both in economic and political spheres, intertwined with one another. They owned most of the agricultural land and around whom the whole village economy revolved. Others were dependent upon them and drew their sustenance by serving the Thakurs. Some sharecropped the land owned by Thakurs, while many others provided labour and services. Majumdar et al. also underlined that though Thakurs showed customary respect to Brahmins, they were treated equally to other castes in everyday life. One of the most interesting aspects of the narrative of Majumdar et al. was the divergent perceptions about the relative social status of castes. For instance, Ahirs gave one account of commensal relations with Thakurs, while the latter contradicted.

Of the two major Dalit castes inhabiting the village, the Chamars were at the bottom of the social and economic hierarchy. Pasis worked closely with Thakurs as their trusted men in roles of messengers, guards, and the like. Chamars were mostly landless agricultural labourers, sharecroppers, and skinned dead animals. *Begar* (unpaid labour) was forcefully extracted from them and was made to perform several tasks besides agricultural labour, including cutting wood and repairing houses. Thakurs ruthlessly imposed conditionalities on sharecroppers belonging to the community. Besides economic exploitation, they were assigned degrading social status and subjected to humiliating behaviour in their daily lives. The practice of untouchability was most stringently practised against them (Majumdar et al., 1955).

However, Majumdar et al. described "functional" relations among castes in a self-sufficient village community, but they also captured the nuances of social change. It all began with resentment among Chamars about the demeaning status and extreme exploitation imposed on them by Thakurs. They started avoiding going to work in their field, even when offered cash wages. Some also took up skinning dead animals as a full-time occupation instead of agricultural labour. They lost social status, even among their community members, but gained economically.

The abolition of Zamindari (intermediaries between the actual tiller and the state) came as a big blow to the complete dominance of the Thakurs. When murmurs about it began, Pasis too started wearing sacred threads (a symbol of twice-born upper castes, Thakur and Brahmin) and avoided receiving *kaccha* food from the Thakurs. Ahirs have always asserted a respectful position before Thakurs, which the latter never explicitly accepted. Several became *Bhagats* (devotees, renouncing polluting and intoxicating habits and food like non-vegetarian and alcohol) and stopped taking *kaccha* food from Thakurs who consumed non-vegetarian food. Similarly, when Chamars began to assert higher status, they refused to eat with Dhobis and Bhaksors, whom they considered lower (Majumdar et al., 1955).

Majumdar's study blasted several prevailing myths about the rural life of the time. First, it revealed the role of power in organising and running the "interdependent" system of reciprocal relationships. The Gohanakallan study made it amply clear that the Thakurs wielded power and used it to enforce norms governing castes. Whenever any group deviated from the rules, it was dealt with an iron fist, not necessarily physical violence all the time but often economic sanctions and social boycotts. For instance, when Pasis, trusted lieutenants of Thakurs, resorted to avoiding *kaccha* food from Thakurs, the latter got sharecropping lands vacated and ensured no other caste served the former. Second, the study underlined tensions debunking the idea of a cohesive village community. In the narrative, angst among Chamars about the humiliation in everyday life and rabid exploitation could be easily discerned. Pasis, too, sought higher status, infuriating the dominant caste.

However, Majumdar et al. sometimes appeared nostalgic about the village community based on reciprocal relationships; they also identified three sites of change: the state, the market, and the community. State intervention in the form of the abolition of zamindari was considered the foremost reason for the downfall of the old village system based on mutual interdependence and solidarity. The Gohanakallan study also underlined migration and closer contact with the urban area as another noteworthy driver of change. For instance, people who visited far-off places returned with stories that village folks listened to with great interest. The location of the village adjacent to the city of Lucknow was also seen as contributing to the transformation. Most importantly, the first driver of change was the urge among people, particularly lower castes, to be treated equally. Unlike most anthropologists of the time, the people at the lower rung of the ladder could easily see the exploitative nature of the old system in the garb of the village community.

Majumdar et al. did not find education to be a major driver of change since it was not accessible to people in any meaningful way. Instead, exposure to the outside world contributed significantly to the process. Ultimately, they believed that new mutual relations would emerge in the interest of the group's survival, sidetracking conflict.

Around six decades later, the First Rural Transformation Survey, 2013 and the Second Rural Transformation Survey, 2020, documented several changes, but some fundamental elements of the village economy and politics persisted. Majumdar et al. noted three hamlets surrounding the main village of Gohanakallan: Malak, Lohungpur, and Balapur. Of these three, Malak was the oldest hamlet recorded by the first survey settlement of 1862. The other two had come up later. Ahirs mainly inhabited these. In the 2013 and 2020 surveys, Malak remained a hamlet of the main village Gohanakallan, and Lohungpur was divided into two villages, I and II. A new urban-style residential settlement named Sainik Nagar had sprung up on the village land. This resulted from the sale of village land to urban people for constructing houses. This was not an exceptional development for villages closer to cities where large-scale diversion of agricultural land was taking place for speculative activities in land and building urban settlements. The unplanned growth of cities had converted the villages' agricultural land, which till recently was the source of livelihood, into "real estate" that could be bought and sold for speculative purposes. This entrenchment of urban capital into rural areas was an economic process and an overhauling of socio-political relations.

The village's population was recorded at 603 in 1955, which increased by nearly five times in six decades. The sex ratio continued to remain low at both points in time. In Majumdar's survey, it was 80 females for every 100 males, ranging between 925 and 940 in the second decade of the 21st century. It improved, but Majumdar's (1955) observation that the sex ratio was inversely related to a community's status in the caste hierarchy did not lose relevance.

In the 1950s, the village did not have a regularly functioning school, and students attended a school at Bakshi Ka Talab, seven kilometres away. By the time of the two rural transformation surveys, one primary and one upper primary school were run by the government in the village. Students still went to Bakshi Ka Talab to obtain further education due to a lack of facilities in the village. Educational expectations have changed immensely. The educational status of the eighth standard neither commanded any respect nor opened new avenues for employment. It was akin to basic literacy without proper schooling in the 1950s. In this relative sense, the educational infrastructure in the village has not made any commendable progress. Similarly, health services in the village have not shown any promising change.

This village was quite distinct from all other villages under the study due to its proximity to the state capital Lucknow, a big city and administrative hub of the government. It has impacted how the village transformed in the last couple of decades. Located just 15–20 kilometres from the district headquarters of Lucknow, expansion of the city usurping the vast area of hinterland not only changed the face of the city but completely transformed the economy and society of villages around every big city in the last two decades. Gohanakallan was not an exception.

Before this, Gohanakallan was just another village. However, not much pluralism existed in the religious composition of the population, with all the resident households following Hinduism, but the caste composition of the village was diverse. Thakurs and Yadavs constituted the bulk of the village population; the former owned the largest area of agricultural land, followed by the latter. The landownership structure of the village had imprints of the legacy of the old zamindari system of the colonial period under which Thakurs held intermediary rights, and Yadavs were recorded tenants. The third segment of the old agrarian relations, the Dalits – both Chamars and Pasis – partly sharecropped land in the personal possession of *zamindars* as insecure tenants and partially worked on their fields as manual labour, continued to own tiny plots at the time of surveys in 2013 and 2020. They were the ones who formed the bulk of the insecure tenants and manual labour, diversifying outside agriculture at a faster pace than expected. However, the agrarian relations in the village have been transformed to a great extent in the last six decades, but the equation between caste and landownership still holds ground.

The village's agriculture was caught in a vicious cycle of low investment, low productivity, and low income from cultivation. Like many other central Uttar Pradesh villages, Gohanakallan kept rotating wheat and paddy. Without a canal or any other public source, electricity-powered tube wells were the primary source of irrigation. However, the primary sources of occupation had shifted away from agriculture. The village's location close to the city provided a perfect backdrop for the emergence of a new section of "property dealers" who essentially played the role of a middleman between urban buyers and farmer sellers of agricultural land. The number of petty traders selling everyday supplies had also swelled over time. A group of contractors and employees would form another new occupation stream. Workers commuting to the city as manual labour, semi-skilled workers, and hawkers also increased during the reference period.

Pandari

The two villages described above were selected since they were studied earlier, providing rich data on a long-term view of transformation. The two other villages, Pandari and Seemli, were randomly selected in 2013 from a list of medium-sized villages with households between 180 and 230. The district selection was purposive, considering the region's representative characteristics, including geography, climate, economic development, and sociocultural features.

Like many other villages of Bundelkhand, Pandari experienced water scarcity during the non-rainy season, despite a canal passing by the side of the village and a pond inside the village that survived on rainwater. Inadequate rainfall and long, scorching summers kept water bodies dry most of the year.

The village of Pandari was 15 kilometres from the district headquarters under the Karvi sub-district and block in the Chitrakoot district of Bundelkhand. The village had no Muslim families and was inhabited by Hindu families. Households belonging to Chamars, Koris, Yadavs, Kahars, Naais, Kalvars, Sahus, Brahmins, and Thakurs lived in the village. Agriculture faced several issues that did not allow the landed class to cultivate two crops. These included the unequal distribution of agricultural land, lack of fertility and water retention capacity in the soil, and the tradition of setting free stray animals.

The farmers only cultivated Rabi crops in the village, including wheat, til, arhar, moong pulse, and jwar. During the Kharif season, most farmers kept their land fallow except for those with irrigation facilities of their own. These farmers cultivated paddy for self-consumption. Because of single-crop agriculture, farmers were developing a feeling of discouragement to continue farming.

Those who did not have cultivable land or had tiny plots sharecropped or leased-in land from landowners. Under general terms, the lessee would bear the expenses of hiring a tractor and thresher, and both parties equally shared the remaining costs. Agricultural produce and by-product were both shared equally. Since this sharecropping system was burdensome for the tenant, people moved away from tenancy.

The village's economy and occupational pattern were constituted around agriculture, with many people engaged as cultivators and agricultural labourers. With increasing stone mining activities in the area, the number of people engaged in the sector and transport has increased. An uptick in trade activities was also visible while animal farming continued. Migration was the only option without any visible dynamism in the local economy. But the labour force of the village, largely unskilled due to the absence of training facilities or opportunities for on-site skill development, migrated to arduous workplaces like brick kilns.

The living standard of the people in this village was relatively low. Over a quarter of the families were drawing drinking water from open wells. They had no private handpumps or access to any nearby public handpumps. Children of these families had limited access to educational facilities since no nearby school was imparting education beyond the eighth standard. Children had to walk three kilometres to reach the nearest school in Shivrampur. The girls' education was neglected entirely in the village under the age-old social belief that women only had to do household chores. This tendency drew strength from the lack of a school in the village, as parents were unwilling to send girls to far-off places.

Most people considered highly educated in the village were hardly High school or Intermediate. The only counted persons in the village were graduates. In this way, one of the reasons for the village's backwardness was the miserable educational infrastructure.

The village was connected to the main road by a *kaccha* lane and a *pakka* street, but no public transport plied on any of these pathways. People living in the village had to cover this distance using private vehicles like cycles and motorbikes. It often created problems for people in urgent need of medical attention. Most of the dwellings in the village were *kaccha*. Lanes and by-lanes in the village have been built without a proper drainage system; streets remain waterlogged. In Bundelkhand, village ponds were built every year for rainwater harvesting. But in this village, no new pond was constructed in the five years preceding the survey. The village was away from basic amenities like a health centre, veterinary centre, regular market, bank branch, post office, bus stop, etc. It was dependent on a neighbouring village, Shivrampur, for many of these facilities.

Most of the village's residents were engaged in manual labour. Still, they were not getting work under MGNREGA. They did not know how many had job cards, nor were they familiar with their rights under the Job Guarantee Act. Except for Brahmins and Thakurs, people from all other *jatis* went to nearby towns like Karvi and Shivrampur to work as headload workers and masons. Some youth also migrated to cities in search of jobs. Women and children of Pandari worked in the harvesting of wheat on fields owned by others and received grains in return. These women said they collected grains for 5–6 months during the wheat harvesting season. Because of this, they had to buy less from the market. Cows, buffaloes, and goats were the preferred animals in the village. Households belonging to almost all castes kept goats. Pigs were reared by one family belonging to the Balmiki *jati*.

The practice of untouchability was prevalent in the village and appeared to regulate people's interactions on many occasions, including at social gatherings. Data collected from Dalits, on the one hand, and corroborated by information from OBCs and non-SC–OBC–ST, on the other hand, revealed that the former received invitations to feasts from the latter but were served food separately. The backwards and UCs also attended Dalits' social functions. But, they avoided eating there unless the food was cooked by a person who belonged to an "acceptable" jati and food was served to them away from the Dalits attending the feast. The practice of purdah (veil) was also widely prevalent in the village. Women were generally confined to their homes and fields. Women rarely went to the market. Only those women whose husbands had migrated out went to the market. Undoubtedly, Pandari was the most backward among all the studied villages on social, economic, and cultural parameters.

Seemli

In the Sadar sub-district of Muzaffarnagar district, the village of Seemli had Hindus and Muslims living side by side. The atmosphere in the village betrayed the recent talk of the district being the hotbed of communal tension. The village had a mosque and a temple where loudspeakers were used for *azaan* and *aarti* without any objection from either side.

Among Muslims, Teli and Saifi, and among Hindus, Jat, Gadariya (Pal), Kashyap, Kahar, Kumhar, and Chamar-Jatav, jatis lived in the village. Muslim Telis of the village traditionally made quilts from cotton, whereas among Saifis, some families worked as ironsmiths and others as carpenters. Both Muslim Lohar and Muslim Badhai used the Saifi caste name. But by the 2020 survey, only 4–5 families were engaged in wood and iron crafts.

Most Muslim households did not own agricultural land. Some of these families had petty businesses like welding workshops or small shops. The majority of people worked as masons, while others as headload workers in the construction industry. On the other hand, among Hindus, most of the Jat households owned land and were engaged in cultivation.

After Jats, the village's main cultivators, Jatavs, were also in agriculture. About 4–5 Gadariya families kept a herd of sheep. However, the reduction of forest area and grazing land posed a great challenge for them. These days, they are grazing sheep on both sides of a railway track. They are being forced to leave their traditional occupation because of these challenges. Elders of the families feel that the next generation will not continue with animal farming.

All the villages under study were centred around agriculture, and the movement of the occupational pattern was towards non-agricultural labour. But the people who traversed this transition make every village a unique story. People were leaving agriculture in Pandari since it was backward and unable to provide for the dependent population. Agriculture was relatively developed in Seemli, but the village's location enabled people, particularly the non-landowning class, to work in the nearby city, raising their income. A simultaneous expansion of trade, transport, and communication facilitated this process while generating new opportunities.

Only Jats, Jatavs, and Pal families had members who got higher education. Some of these people had obtained PhD, MBA, B Ed, MA, and undergraduate degrees. Members of other jatis had hardly received education up to the fifth and eighth grades. The village school imparted education only up to the eighth standard, forcing Muslim girls to leave education at that point. Some of these families sent their boys to study beyond the eighth standard, but they did not favour sending girls outside the village for schooling. They would have allowed girls to study beyond the eighth standard if a school had been established in the village. Many Muslim families found it futile for the boys to study as they had to work in the same trade even after obtaining an education. They preferred skill training that would enable them to be absorbed early in the job market.

The highest number of people with government jobs in the village were from the Jatavs. After that, some people from Jat and Pal jatis also had government jobs. Most people in this village were associated with agriculture except those who did not own land. They worked as manual labour in Muzaffarnagar city near Seemli, facilitating commuting to work from home.

Daily wages for casual labour were relatively high in the district. At the time of the survey, towards the end of 2020, the daily wage of a mason was around Rs 600, and that of a helper was from Rs 400 to Rs 450. Agricultural workers earned between Rs 350 and Rs 400 for a day's work in the fields, while factory workers could fetch Rs 500 to Rs 600 as daily wages. There were many iron rod manufacturing units where manual labourers from Seemli found work in loading, cutting, and bending iron rods. Regular factory workers got a monthly salary of Rs 15000 to Rs 20000. Some people from the village worked at a welding workshop to earn a daily wage of Rs 400 to Rs 500.

Owing to the proximity of sugar mills, the largest crop of the village was sugarcane, which covered more than three-fifths of the area under cultivation. At the same time, wheat and chari were cultivated in the remaining area for self-consumption and animal feed, respectively. All the families associated with agriculture also kept milch cattle. Most domestic animals were of a superior breed, and owners cared for them very well. However, there was no milk cooperative in the village at the time of the survey, but local vendors collected milk from the producers at Rs 40 for buffalo milk and Rs 30 for cow milk and sold it in the market in Muzaffarnagar. Some households that did not own agricultural land were engaged in milk production and supply in the village. Milk was also consumed by families residing in the village. Women played a more significant role in decision-making and providing labour for animal farming.

The village received water for irrigation from a canal flowing adjacent to it, a public tube well, and numerous private tube wells. The availability of this infrastructure was hardly a surprise, as western Uttar Pradesh has a long history of possessing it. This factor was one of the primary reasons for targeting these areas for the "green revolution". But in the case of this village, the necessity of building a canal arose only when the river passing through the village got polluted, and its water became unfit for irrigation. The same goes for drinking water. The river water served the village well until industrial waste polluted it, and people shifted to private and government handpumps.

The village has been electrified, and streetlamps have been installed on electricity poles. A public water cooler has also been affixed near the residence of Gram Pradhan (Chairperson, Village Panchayat). There was a Syndicate Bank branch in the village. Despite these developments, Seemli had only one upper primary government school and a private school. Families sent their wards to Muzaffarnagar town for better and higher education if they could afford to do so. Most of the families understood the importance of education in moving forward. The urban middle-class aspirations have caught the imagination of everyone in the countryside, particularly those well-off in the rural economy. They invested in education, knowing it was a vehicle for realising the urban dream. The village also had no primary health centre, and residents depended on the city to access health services.

Unlike other villages under study, the practice of untouchability was relatively weaker in commensal relations in Seemli. People belonging to different castes ate at each other's places.

Concluding Remarks

A simple story like this could demystify the essentialist framework of looking at the rural. It reiterates that every village has its own sociological reality. The narrative also showed that the landownership structure and the interests of the dominant landowners could trigger a change. The four villages had two different types of caste-landownership patterns. The upper castes (UC) owned larger tracts of land in Senapur, Gohanakallan, and Pandari, while Seemli landowners belonged to a peasant caste. The agricultural economy was relatively underdeveloped in the first two villages located in the eastern and central regions, whereas it was developed in the western region village and backward in the Bundelkhand village.

The growth engine appeared to be the dominant landowning class, not necessarily a conventional large farmer but relatively bigger landowners in the village's landownership structure. In villages where this class of farmers invests in agriculture to produce for the market, it leads to the overall development of agriculture. These farmers were often responsible for "tractorisation" of agriculture, acquiring and renting farm implements, initiating cooperatives dealing in inputs and outputs, and opening marketing channels, transforming the village's agriculture.

However, this class of farmers would not behave like capitalist farmers everywhere. Two models emerged on agricultural incomes and interest in developing agriculture. The peasant landowners of Seemli have been investing heavily in raising the productivity of agriculture and diversifying beyond food grains. The landowners of the other three villages, dominated by UCs, have paid scant regard to the agricultural enterprise. The qualitative data revealed that two major factors, among others, influenced the landowning classes' investment interests. They often evaluated the potential of raising income from agriculture and the possibilities of non-agricultural incomes. Several other factors came into play, including natural endowment, the availability and accessibility of the market, and the social development of the population.

For instance, the large landowners of Senapur might have realised the limited potential of agriculture due to the problem of waterlogging in low-lying areas of the village. They focused on education rather than agriculture, leading to the setting up government-run schools under political pressure from the community. Many UC landowners preferred to migrate to white-collar employment in post-independence India. Pandari's UC landowners weren't equally lucky. While they faced nature's fury in water scarcity and low soil fertility, they were also excluded from educational opportunities. The village's remote location in a backward area ensured that the entire population had no option but to migrate to arduous jobs without skill training. The lack of public irrigation infrastructure obstructed agricultural development in Gohanakallan, too. The market access in the nearby city alone could not enthuse entrepreneurship among the village agriculturalists. The village's location in an economically developed region did not help either. By contrast, abundant irrigation infrastructure facilitated a shift from food grains to cash crops in Seemli.

Though UCs have, by and large, shrugged off the tag of non-cultivator owners in the ever-increasing mechanisation of agriculture, the productivity of peasant caste-owned farms has mostly been higher. It was on account of the level of engagement between both groups of farmers. While the peasant landowner families routinely engaged in manual labour in the fields, the self-cultivation of UCs has been confined to machine operation and managerial roles. In other words, armed with machines and family labour, peasant castes outwit UCs who relied only on machines. Not only that, but in many villages, direct descendants of ex-landlords were still hesitant to take up self-cultivation.

The class of large landowners may or may not push for the transformation of agriculture, but a thrust from below towards rural change was inevitable. Change-makers were mostly understood to be active participants, but abandonment could be equally effective. It could be particularly devastating for a structure if the base starts shifting. Despite many variations among studied villages, a large labour force consisting of Dalits was common. It was upon these people's labour that rural societies were built. The rural Dalit working class has been yearning for liberation from the repressive agrarian structure whose influence extended beyond the economic sphere. The opportunity boom in the urban economy came as a breather for them. Working conditions on a construction site might have been harsh, but caste anonymity was a much-awaited relief.

The rural working class might have preferred to get absorbed in the urban economy, but instead, they were kept on the periphery, a dispensable labour force. No doubt, the chimera of decent urban life has caught the imagination of all rural folks, but the extremely exclusionary urban economy has its ways of dealing with low-skilled migrant workers. Insecure jobs and insufficient earnings meant seeking a livelihood, whichever way it came. The economic aspects of circular migration have been discussed widely, primarily underlining the diminishing carrying capacity of rural farm and non-farm sectors and opening opportunities in burgeoning urban industries, particularly construction, workshop manufacturing, and household services.

But it played a crucial role in rural transformation, much more than permanent migration. The circular migrants returned with urban lifestyles and accumulated earnings from urban jobs. An uptick in demand for industrial goods and services in a perpetually cash-strapped rural economy reflected this tendency. The other much-desired change was the weakening rural power structure based on caste-land congruence due to Dalits' decreasing dependence on agricultural wages and land tenancy. It would be naïve to believe that economic changes alone damaged the age-old power structure. Political changes preceded it. For instance, cracks in the Senapur fortress of the UC landlords had started appearing during village local body elections soon after the country's independence. Similarly, the underclass of Gohanakallan had started asserting autonomy from the complete domination of landlords when *Zamindari* abolition was set in motion.

Social science research has extensively analysed the state's role in backward regions as the leading agency in pushing the transformation agenda (Athreya et al., 1990). State initiatives came in the form of transportation, power, and social development infrastructure. But not much attention has been given to the interplay of various classes on the site of the state in prioritising initiatives. For instance, transport and communication infrastructure has grown tremendously in the last few decades. But the social development infrastructure has not witnessed equally impressive growth, and wherever its size has grown, the quality has been missing.

The growth of transportation, communication, and financial infrastructure has facilitated rural linkages with the urban in labour, commodity, and finance. All the studied villages have been connected by all-weather roads that have contributed majorly to the seamless circular migration. The spread of mobile telephony connected construction workers with their prospective employers and labour contractors. They didn't have to wait for an opportunity sitting in the metropolis. Instead, they were called as and when the occasion demanded. When there was no work, they returned home. This was the only way they could survive with meagre wages against high city living costs. Workers employed in seasonal manufacturing and service industries also followed similar patterns. The wide gap between urban survival costs and wages also compelled most worker families to stay in villages. Unlike yesteryear's migrant workers who visited native places once in a couple of years, the new migrants maintained their ties with agriculture by monitoring it over the phone and participating directly. Qualitative data revealed that decisions on women-managed farms were often made by the "head of the family" staying in a metropolis.

The phenomenon of circulatory migration has been scaling new heights, riding on the back of improved infrastructure. The other consequence was flooding rural markets with industrial goods, leaving them indistinguishable from their urban counterparts. Some villages might be "self-sufficient" in agricultural commodities in terms of production, if not distribution; they consume industrially produced commodities at varying prices depending on their paying capacity. The market caters to far-flung rural areas with ease, taking advantage of connectivity.

However, the public in the studied villages appeared content with the improved transport, communication, and financial infrastructure, but working-class people were also concerned about the quality of education imparted in government schools and the lack of proper medical facilities in public health centres. Those who could pay sent their wards to private schools in the village or nearby town and sometimes to the city. They also availed themselves of health services from private practitioners.

The state-initiated rural transformation appeared to prioritise infrastructure where private investment was not forthcoming. Due to the scale of investment and the negligible possibility of commensurate returns, village roads weren't a good investment option. This strategy left space for private capital in the social development sector. One might view it from the prism of private and public investment complementarity. But the question arises of de-prioritising those sectors of the economy for public investment where only the poor were dependent on it. The rich could buy education and health services from the market, making a readily available opportunity for private investment. This scheme of public roads and private schools entirely serves the interests of the rich, but the poor lose out on social development. The site of state intervention in rural transformation is quite interesting in revealing the interplay of capital and rural classes. This also reveals the dynamics of dominant interests becoming "public good".

The state intervention was not gender-neutral either. The recent fall in women's workforce participation might have been attributed to growing prosperity among rural households, but the nature of emerging jobs and the preparedness of women workers in training and spatial distance often skipped the attention of social sciences. Women's jobs were lost in the mechanisation of agriculture and a shift from farm and household economic activities to off-farm and industrial economies. It was not a natural course of action but a socially mediated process. The loss in productive capacities of the rural and the concentration of economic activities in the urban wasn't the only path for rural transformation. The socially embedded nature of the new technology displaced women from their productive roles and consigned them to unwaged and unpaid reproductive roles. The state facilitated an exclusionary transformation agenda, witnessing massive pushback against women that was wrongly termed as "withdrawal" in literature.

Dalits might be breathing a sigh of relief for emancipation from the overbearing agrarian and caste structure. Some cash earnings from urban jobs were also helpful. But they hardly experienced vertical mobility. Agricultural workers and marginal farmers in almost all the studied villages frequently became manual labourers in industries. The state expedited their physical movement but hardly took an interest in their vertical movement by deprioritising education and workers' social security. The transition of the rural underbelly to urban footloose migrants was the characteristic feature of Indian rural transformation.

3 Inequalities of Opportunities and Outcomes

Demography, Education, and Assets

Based on longitudinal data, this chapter emphasises that demographic changes pushed the transformation of rural areas, and access to quality education shaped the dynamics of change, contributing to inequality among social groups. The fall in fertility rate over a period reflected in a dipping share of the child population, and a slight uptick in the elderly population culminated in a broadening base of the working-age population in rural areas. This inadvertently provided an asset for rural to negotiate with urban. But rural, particularly their underbelly, failed by the state and the market in arming this working population with quality education, found themselves as "cheap labour" for the burgeoning economy in the cities. Only legacy asset owners in rural could break into the elite club of white-collar job holders and business owners. The rising qualitative asset inequality in rural areas was only the tip of the iceberg since a larger share of professionals settled down in towns compared to manual workers who formed the bottom of the hierarchy in both locations.

Demographic Transition

The country has been undergoing a demographic transition during the last three decades, reflected in a falling population growth rate since 1991. But largely, the debate was hijacked by the economic consequences of fast demographic changes rather than its implications for social relations, particularly caste and gender. The strand of discourse on social relations believed that demographic changes would ameliorate gender relations if not completely transform them. The second demographic transition theory believed in greater autonomy for women and a weakening of the gendered division of labour, with men as earners and women as caregivers. Demographic changes were also expected to impact women's negotiating power in family and parenting positively. However, none of these claims was without contest (James, 2010).

Against the high hopes of a gender-just demographic transition, state control over women's reproductive function seeking the interests of the market

DOI: 10.4324/9781003588245-3

persevered. Population policies in countries such as China, where a rigid one-child policy was imposed in the 1980s, and the subsequent relaxation or population stabilisation and family planning programme in India, have been premised on manipulative tendencies rather than recognising women's agency. These policies prioritised the social reproduction of labour at a required level, not hesitating to infringe on women's autonomy and often going to the extent of coercing them. These propensities in the policies have endured in the face of mounting evidence that gender-sensitive social development initiatives would go a long way in addressing demographic challenges. Among the economic instrumentality that has pushed the demographic policies so far was the subsidisation of labour for capital expansion. Even in a populated country like India, getting the required number of able-bodied workers, sans the burden on capital to skill them, and almost without any bargaining power and retention cost, drove policy agenda on demography (Bagchi, 2004; Chandrasekhar et al., 2006). This tendency of capital vis-à-vis social reproduction of labour has colossal consequences for rural transformation in contemporary times. As we see from the data in this study, the rural emerged as the site of reproduction while the urban retained the title of the site of production. And these rural workers without basic training formed the footloose migrants deployed in the most basic manual work, returning home when shunted out at will.

Coming to the population data, Uttar Pradesh's population growth rate has been higher than the national average. Among the studied districts, Lucknow and Chitrakoot had a growth rate of 2.5% and above between 2001 and 2011. However, the reasons for the growth rates might be different. Lucknow is a destination for immigrants from several parts of the state. Going by Census 2011 figures, the state had a lower than national average urbanisation at 22 per cent. The disaggregated data presented a mixed bag of evidence. One of the most critical indicators of gender equality – the sex ratio among juveniles – has worsened in many places. But at the same time, the gender difference in literacy rates has come down slightly, similar to infant mortality and life expectancy. This apparent contradictory evidence demands further probing into the transformation of social relations during demographic changes (James, 2010).

Demographic Changes in Studied Villages

In the methodology chapter, it has been mentioned that the panel of households surveyed in 2013 and 2020 differ to an extent on account of three factors. First, some households experienced a split during this interim period. Second, some new households left out in 2013 were surveyed in 2020. Third, some households surveyed in 2013 could not be cross-matched with households surveyed in 2020. All the results in this chapter and subsequent chapters are based on panel data of 1348 households visited twice in 2013 and 2020.

Age-group-wise distribution of selected villages in 2013 is presented in Table 3.1. The 2013 survey estimated that up to 14 years of the population was 34.5 per cent of the people in Senapur. The share of the elderly population (60+) was 11.2 per cent then. Seemli had a share of up to 14 years population at 34.8 per cent and above 60 at 7 per cent. The remaining 58.2 per cent of the population was in the age bracket of 15–59 years.

Similarly, close to 38 per cent of the population in Pandari was up to 14 years of age in 2013. The share of the above 60-year-old population was 8.8 per cent. And the population between the two cohorts was 53.1 per cent of the total population of the village. In 2013, Gohanakallan had a 34.8 per cent share of the population aged up to 14 years, while 8.6 per cent of the population was aged above 60 years. The percentage of the population between 15 years and 59 years was 56.5 per cent.

In Senapur, the population up to 14 years was 26.30 per cent in 2020, while 11.89 per cent of the population was above 60 years of age. The remaining 61.81 per cent of the population was in the age bracket of 15–59 years. Seemli had a relatively larger younger cohort of up to 14 years at around 28 per cent, while a much smaller elderly population share of 8.4 per cent. However, these villages have a comparable proportion of the population aged 15–59 years at 62–63 per cent. Over 31 per cent of Pandari's population was up to 14 years, while 9.45 per cent was over 60 years. At the same time, 59.5 per cent population of the village was in the age bracket of 15–59 years. The share of the young cohort of up to 14 was estimated in Gohanakallan at 26.52 per cent. It is the village with the second-lowest proportion of the above 60 years population at 8.84 per cent. The number of people in the age bracket of 15–59 years was 64.6 per cent.

When juxtaposing the data from the 2013 survey with the 2020 survey, it became very clear that the share of the population up to 14 years in all four villages was decreasing gradually. Taking the people of all four villages together, the up to 14 cohorts had a share of 35.4 per cent of the population in 2013, reaching 27.6 per cent in 2020. Senapur's percentage of up to 14 cohorts decreased from 34.5 in 2013 to 26.3 in 2020. In Seemli, it went down to 28.2 from 34.8, in Pandari from 38 to 31, and in Gohanakallan from 34.8 to 26.5. The pattern was keeping in with the national trend of a decreasing share of the juvenile cohort, which was considered a fallout of the falling fertility rate in the previous decade.

The same data also revealed a slight increase in the share of the 60-plus population from 9 per cent to 9.8 per cent of the total population. In Senapur, the percentage of the 60-plus population increased from 11.2 per cent to 11.9 per cent, in Seemli from 7 per cent to 8.4 per cent, in Pandari from 8.8 per cent to 9.5 per cent, and in Gohanakallan from 8.7 per cent to 8.8 per cent.

In all four villages, the share of the population up to 14 years old has gone down by seven to eight percentage points. But a hasty conclusion about population growth in a village would be premature since several factors need to be considered, including in-migration and out-migration. The two-point

Table 3.1 Population Distribution in Different Age Groups: 2013 and 2020

Age-Cohort	*0–14*		*15–29*		*30–59*		*60+*		*Total*	
Village	2013	2020	2013	2020	2013	2020	2013	2020	2013	2020
Senapur	34.52	26.30	27.19	29.28	27.14	32.53	11.16	11.89	100.0	100.0
Seemli	34.79	28.15	30.02	31.30	28.16	32.17	7.02	8.38	100.0	100.0
Pandari	38.06	31.03	26.49	29.00	26.63	30.52	8.83	9.45	100.0	100.0
Gohanakallan	34.81	26.52	27.46	33.29	29.08	31.34	8.65	8.84	100.0	100.0
Total	35.35	27.64	27.71	30.85	27.88	31.70	9.06	9.81	100.0	100.0

Source: UP Rural Transformation Survey (2013, 2020).

survey data also revealed that the population in all four villages was getting younger. In all villages, the share of people aged 60 plus in the total population had increased by around one percentage point between 2013 and 2020. But the working-age population (15–59 years) increased by around seven percentage points.

Table 3.2 gives the social group-wise composition of the population in all four studied villages in 2013. The people in Senapur comprised 55.5 per cent SCs, 31.6 per cent OBCs, and 12.7 per cent non-SC–OBC–ST group. In Seemli, the share of the OBC population was 59.2 per cent, and the SC population constituted 40.2 per cent. The distribution of the social group-wise population in Pandari was 44.4 per cent, 42.6 per cent, and 12.7 per cent, respectively, for SC, OBC, and "general" categories. Gohanakallan's population comprised 52.8 per cent OBCs, 36 per cent SCs, and 11 per cent non-SC–ST–OBC.

In 2020, the social group-wise composition of the villages had changed marginally. Taking the population of all four villages, the SC population was 43.94 per cent in 2013; it became 44.42 per cent in 2020. The share of the OBC population was 46.2 per cent in 2013; by 2020, it came down to 45 per cent. During the reference period, the percentage of the non-OBC–SC–ST population slightly went up from 9.6 per cent to 10.3 per cent.

The religious group-wise composition of Senapur was 94 per cent Hindus and 5.7 per cent Muslims. In Seemli, it was 77 per cent and 23 per cent, respectively, for Hindus and Muslims. In the remaining two villages, all the residents reported their religion to be Hinduism. The share of Muslims in the population of Senapur came down to 5.5 per cent in 2020, and the percentage of Hindus went up marginally to 94.5 per cent. Likewise, the share of the Hindu population in Seemli went up to 78.3 per cent, and the share of the Muslim population came down to 21.7 per cent in 2020.

The total population in Senapur grew by about 2 per cent per annum during the seven and a half years between the two surveys. The preponderant group of the village, SC, grew by an average population growth rate of 2 per cent per annum. The growth rate of OBC was 1.5 per cent, while the non-OBC–SC–ST group population grew by around 2.8 per cent. It could also be due to the low base effect. Seemli's population saw a negative growth rate during the reference period, with the SC group showing a marginal growth rate of 0.13 per cent per annum and the OBC population revealing a negative growth rate of 0.32 per cent per annum. Pandari experienced a population growth rate of about 1 per cent per annum. SC and OBC groups grew almost equivalent to the village growth rate, while the non-OBC–SC–ST population grew at 0.63 per cent per annum (Table 3.3).

The Hindu population in Senapur grew at an annual rate of 2 per cent, while the Muslim population in the same village registered a growth rate of 1.6 per cent during the seven and a half years between the two surveys. In two other villages, only one religious community resided.

The fastest-growing age cohort in all four villages was 30–59 years. In Senapur, it grew at the rate of 4.5 per annum, followed by 2.9 per cent in

Table 3.2 Population Distribution in Different Social Groups: 2013 and 2020

Social Group	*SC*		*ST*		*OBC*		*General*		*Total*	
Village	2013	2020	2013	2020	2013	2020	2013	2020	2013	2020
Senapur	55.48	55.70	0.28	0.33	31.57	30.50	12.67	13.48	100.0	100.0
Seemli	40.29	41.15	0.46	0.34	59.24	58.51	0.00	0.00	100.0	100.0
Pandari	44.36	44.61	0.41	0.38	42.58	42.70	12.65	12.31	100.0	100.0
Gohanakallan	36.00	35.21	0.24	0.16	52.84	52.65	10.92	11.99	100.0	100.0
Total	43.94	44.42	0.33	0.29	46.17	45.02	9.56	10.28	100.0	100.0

Source: UP Rural Transformation Survey (2013, 2020).

Table 3.3 Social Group-wise Annual Growth Rate of Population between 2013 and 2020

Village	*SC*	*ST*	*OBC*	*General*	*Total*
Senapur	2.04	3.84	1.53	2.82	1.99
Seemli	0.13	-4.48	-0.32	0.00	-0.15
Pandari	1.07	0.00	1.03	0.63	0.99
Gohanakallan	-0.05	-5.39	0.20	1.50	0.25
Total	0.97	-1.11	0.49	1.80	0.83

Source: UP Rural Transformation Survey (2013, 2020).

Table 3.4 Age-group-wise Annual Growth Rate of the Population between 2013 and 2020

Village	*0–14*	*15–29*	*30–59*	*60+*	*Total*
Senapur	-1.62	3.02	4.51	2.88	2.01
Seemli	-2.93	0.41	1.64	2.22	-0.15
Pandari	-1.70	2.24	2.87	1.94	1.02
Gohanakallan	-3.31	2.87	1.27	0.55	0.26
Total	-2.42	2.30	2.58	1.91	0.84

Source: UP Rural Transformation Survey (2013, 2020).

Pandari, 1.6 per cent in Seemli, and 0.6 per cent in Gohanakallan (Table 3.4). Another interesting point from the data was the negative growth rate registered by the under-14 group in all four villages. In all four villages taken together, it was 2.4 per cent. The steepest fall in the growth rate of the below 14 years age cohort was recorded in Gohanakallan at –3.3 per cent, followed by –2.9 per cent in Seemli, –1.7 per cent in Pandari, and –1.62 per cent in Senapur.

One of the most important indicators of the demographic health of society is the sex ratio. All four villages revealed similarities in many other aspects of demography, but when it came to the sex ratio, they differed. Senapur, located in eastern Uttar Pradesh, revealed a relatively high sex ratio of 996 females per 1000 males in 2013. Gohanakallan, in the central part of the state, recorded a sex ratio of 942, followed by Pandari in Bundelkhand at 909 and Seemli in western UP at 792. In 2020, Gohanakallan registered a sex ratio of 991, followed by Senapur at 885, Seemli at 826, and Pandari at 814. At the village level, significant variations were seen in the sex ratio, but taking the population of all four villages together, the sex ratio was 918 in 2013, reaching 891 in 2020.

Activity Status

Table 3.5 presents the activity status of the entire population of all four studied villages at two points – 2013 and 2020. The share of the child population

Table 3.5 Population Distribution by Activity Status: 2013 and 2020

Activity Status	*Senapur*		*Seemli*		*Pandari*		*Gohanakallan*		*Total*	
	2013	2020	2013	2020	2013	2020	2013	2020	2013	2020
Child	11.56	7.92	9.42	8.37	14.27	10.62	9.67	7.61	11.03	8.43
Student	32.86	29.70	34.75	30.70	30.18	27.91	32.66	30.72	32.65	29.85
Working	32.90	33.54	39.19	34.89	43.69	33.95	40.25	32.95	38.66	33.68
Household Activities	16.48	22.20	13.46	21.39	6.65	22.12	12.79	23.94	12.76	22.58
Unemployed	0.91	1.22	0.13	1.21	0.34	0.95	0.32	0.56	0.45	0.96
Retired/Pensioner	0.76	0.82	0.07	0.13	0.27	0.38	0.41	0.56	0.41	0.52
Old/Handicapped	4.49	4.28	2.98	2.90	3.22	3.75	3.29	3.39	3.55	3.64
Others	0.05	0.33	0.00	0.40	1.37	0.32	0.61	0.28	0.48	0.32
All	100.00	100.0	100.0	100.0	100.00	100.0	100.00	100.0	100.00	100.0

Source: UP Rural Transformation Survey (2013, 2020).

was 11.6 per cent in Senapur in 2013. Students and the working population had an equal share in the total population at 32.9 per cent each, while 16.5 per cent of members of the households reported household activities. Other categories like old/handicapped, unemployed, and retired had relatively infrequent reporting.

The activity status pattern of Seemli was also not significantly different, with 34.8 per cent of students, 39.2 per cent of the working population, and 13.5 per cent engaged in household activities. Children constituted 9.4 per cent of the village population. In Pandari, the working population was relatively higher at 43.7 per cent while reporting household activity at a low level of 6.7 per cent. Students and children constituted another 30.2 per cent and 14.3 per cent, respectively. Gohanakallan's pattern of the activity status of the population resembled that of Seemli, with 40.2 per cent working population, 12.8 per cent of household activities, 32.7 per cent of students, and 9.7 per cent of children.

In 2020, while the share of students and the children population came down by three percentage points each, the reporting of household activities went up by about six percentage points. The working population almost maintained its share of the total population of Senapur. In Seemli, the reporting of the student and working populations together came down by over 8 per cent while the share of household activities rose by a similar magnitude.

The highest jump in the reporting of household activities was seen in Pandari, where it jumped to 22.1 per cent in 2020 from 6.7 per cent in 2013. In this village, a massive restructuring of the activity status of the population had taken place during the reference period, which was reflected in a close to 10 per cent fall in the share of the working population, a 2.5 per cent fall in the percentage of students, and 3.5 per cent fall in the share of the child population. In Gohanakallan, the share of the population reporting engagement with household activities rose 11 percentage points, while the share of the working population came down by 7.3 percentage points. The student and child population share also decreased by two percentage points each.

Taking all four villages together, the share of the population reporting engagement in household activities rose by ten percentage points. This rise was at the cost of reporting on work, students, and the child population.

Among social groups, Dalits had a minor share of household members reporting engagement in household activities at 11 per cent on average in all four studied villages, compared to 13.7 per cent of OBCs and 16 per cent of the non-ST–OBC–SC group in the first survey. The share of household members engaged in household activities was related to their location on the caste hierarchy. Another noteworthy point was the increasing share of this category in all three social groups by around ten percentage points in 2020 compared to 2013. Both these observations, the relationship of the share of population engaged in household activities with caste hierarchy and the expansion of this category, were significant and needed further probing.

The mystery of transformation in activity status could only be unpacked by considering gender-wise disaggregated data. While males at both points in time had a negligible presence in household activities, it was only because of the massive swing in the share of females, from an average of 26 per cent in 2013 to 47 per cent in 2020, that caused the swelling of this category. During the same period, female engagement in "working" decreased from 28 per cent to 9.4 per cent. Undoubtedly, this period witnessed a massive withdrawal of women from the workforce to household chores alongside an increase in the share of males in the working category from 48.7 per cent to 55.3 per cent. Surprisingly, this phenomenon was present in all four villages, despite differences among them on several other indicators.

Education and Transformation

This section begins with two footnotes about the literature on rural transformation. First, demographic changes continue to form an important analytical aspect of social change in rural India. They are considered a crucial contributor to the process of transformation. Second, education has increasingly found only a passing reference in these studies, and even if it found adequate focus, it was not built into the transformation analysis. These studies did not consider the role of education as a major contributor to social change. Wherever the impact of education was assessed, it was in the form of opportunities in urban areas and income. However, no study foregrounded this point, but the organic impact of education on social change was assumed to be minimal at best (Kurian, 1983; Lieten and Srivastava, 1999; Himanshu et al., 2016; Jodhka, 2019; Otten and Simpson, 2016).

However, education figured in the rural and urban equation in numerous other ways where the former was prodded to become the latter. These forms were crucial for rural transformation debates. The first was a pedagogic role that emerged with the introduction of modern education during colonial rule, argued Kumar (2014), which ended up placing urban on the pedestal of rural's teachers, reminiscing about colonial masters assuming the role of an adult civiliser vis-à-vis the native, a child. Second, the city was where the village was imagined and reimagined in creating and storing knowledge about it. The village was known to the outside world through urbanites who investigated the village primarily from a utilitarian perspective (Kumar, 2014).

In this rural and urban dynamic, education was assigned an instrumental role in preparing the inhabitants of the former for mobility towards the latter. Despite putting the intrinsic value of learning in the backseat, education could open the doors to economic opportunities and social status ascendency in urban areas. Modern education prepared people to live in towns rather than villages, and it was the primary medium through which rural folks internalised the city–village relationship. This established a spontaneous connection between education and urban areas and disconnected it from rural areas (Kumar, 2014). Rural transformation studies adequately focus on white-collar migration but often overlook the delicate link of education with this process.

While preparedness for and an attraction to urban life driven by education brought people to towns, the inhospitable terrain of villages for educated people was no less imperative. The earlier notion of educated people being the torchbearers of values and tradition was radically transformed with modern education, supposedly based on social change. However, the relationship between education and modernity was not as straightforward; nonetheless, it produced modern aspirations among them and inculcated capacities for participation in democratic processes. The state invested in imparting basic literacy among hitherto excluded people, as visible in data from village studies. Education was an indicator of transformation and contributed to its meaning (Kumar, 2014).

Education and transformation interface with each other in two forms. The first is related to what is indicated by Einstein's famous quote that education is what remained when everything learnt at school is forgotten. Humankind's quest to develop an understanding of the world and oneself has always been an integral part of the philosophy of education (Dewey, 1915). The human development approach defined them as fostering human capabilities, critical thinking, and imagination. This perspective seeks to inculcate three values for decent global citizenship. First, citizens should be capable of self-criticism and the ability to see through their own traditions, using their thought processes rather than succumbing to authority. Second, the modern democratic citizen should be adept at seeing herself as a member of a diverse society, familiar with the culture and history of cohabiting groups. Education should also develop the capability of "narrative imagination" in a citizen that allows one to place oneself in a different person's place and understand her emotions and aspirations (Nussbaum, 2008). Disappointingly, this function of education has received scant attention from rural transformation studies.

The other function of education was related to income and occupational mobility. This was no small contribution to transformation and vice-versa. Unni (2001) found a strong correlation between educational achievement, the probability of landing a salaried job, and income levels. The study based on NCAER-HDI data from rural India revealed that any education helped gain access to salaried jobs that almost ensured survival above the poverty line. But the threshold level for STs was primary schooling, while for SCs and Muslims, it was middle schooling, at which entry into salaried jobs became significant. Unni found a significant positive correlation between the level of education and returns to education among all ethnic groups, except for Muslim men in salaried jobs. Despite the lowest mean years of schooling, SCs had higher returns to education, possibly due to positive discrimination policies. Education also enabled the diversified choice of occupation (Unni, 2001).

It is this emphasis on the instrumentality of education that discourse on education and transformation did not go beyond access to educational institutions. What happened in those schools often evaded the attention of these works. Nambissan (2009) argued that the rampant practice of caste discrimination in schools robs students of different castes of the only avenue in a hierarchical society where they could share meals and participate in activities on an equal footing. Nambissan identified drinking water and food as the most potent sites where the practice of untouchability was widely practised, often with the tacit approval of the teachers. Besides, prayer was another place where Dalit students felt discriminated against. They did not participate equally in co-curricular activities and were often treated harshly in classrooms. When distributing the responsibilities to students during school functions, Dalit students were often given menial work like sweeping while the so-called upper caste students served water and snacks to guests and teachers. The caste identity of the Dalit students was invariably pointed out of context (Nambissan, 2009). The persistence of these practices surely puts a question mark on the kind of transformation rural India was experiencing.

Modern education was expected to prepare critical-thinking citizens, aid income and occupational mobility, and eradicate caste and gender discrimination. Theoretically, "transformation", as discussed in Chapter 1, subsumed the perspective of Nussbaum, the formulations of Unni, and confronted the failures of the education system identified by Nambissan. But it would be to overstressing the obvious that things have rolled out differently in third-world countries. It might be stressful as it is, but this also leaves much space to probe the co-constitutive interface of education and transformation in rural India. While not denying the huge import of the quality of education, the data presented in the chapter is limited to probing access to education.

Educational Status

Like the demography of the village, the educational status of the village population has experienced a noticeable change. The share of the illiterate population to the total population in all four villages has come down from 33.8 per cent to 30.20 per cent. In Senapur, the proportion of people reported to be illiterate in 2013 was 32.16 per cent, which came down to 24 per cent in 2020 (Table 3.6). In Seemli, this proportion marginally increased from 28.04 per cent to 29.9 per cent. Pandari made healthy progress in reducing the share of the illiterate population from 40.84 to 34.9 per cent. Comparable figures for Gohanakallan were 35 per cent and 33.3 per cent in 2013 and 2020, respectively.

The share of the population of all four villages that completed intermediate was 5.52 per cent in 2013, which climbed to 9.6 per cent in 2020. In Senapur, this increase was 4.2 percentage points, followed by Gohanakallan at 4.2 percentage points, Pandari at 3.57 percentage points, and Seemli at 3.5 percentage points.

Similarly, there has been a steady improvement in the proportion of the population who had attained graduation in all four villages, from 3.06 per

Table 3.6 Population Distribution by Educational Status: 2013 and 2020

Educational Status	*Senapur*		*Seemli*		*Pandari*		*Gohanakallan*		*Total*	
	2013	2020	2013	2020	2013	2020	2013	2020	2013	2020
Illiterate	32.04	24.04	27.97	29.90	40.31	34.92	35.01	33.33	33.80	30.19
Below Primary	15.88	12.58	18.56	11.76	12.59	11.49	14.66	10.67	15.38	11.61
Primary	14.45	13.12	17.89	16.54	17.73	15.34	17.99	14.50	16.93	14.62
Middle	12.46	13.91	15.57	17.51	14.72	18.93	16.69	17.12	14.90	16.57
High school	10.33	13.25	9.68	9.13	8.01	7.06	7.96	9.36	8.98	10.05
Intermediate	8.06	12.21	5.90	9.41	3.56	7.51	4.26	8.48	5.52	9.59
Graduate	4.69	8.18	2.92	4.36	2.05	4.11	2.36	5.85	3.06	5.95
Postgraduate	1.80	2.70	1.19	1.38	0.34	0.64	0.61	0.68	1.01	1.41
Total	100.00	100.0	100.0	100.0	100.00	100.0	100.00	100.0	100.00	100.0

Source: UP Rural Transformation Survey (2013, 2020).

cent in 2013 to 5.95 per cent in 2020. During these seven years, the share of the graduate population in Senapur had gone up from 4.69 per cent in 2013 to 8.18 per cent in 2020, 2.92 per cent to 4.36 per cent in Seemli, 2.05 per cent to 4.11 per cent in Pandari, and 2.36 per cent to 5.85 per cent in Gohanakallan.

Social group-wise disaggregated analysis gave an interesting pattern of change in educational status in the studied villages. The share of illiterates among Dalits was the highest in all four villages, but a drop in the share of illiterates was also steepest in this social group. During the seven years, on average, Dalits recorded a fall of seven percentage points in the share of illiterates from 38 per cent to 31.4 per cent, while a similar fall for OBCs and the non-OBC–SC–ST group was around one percentage point each. Despite this effort to catch up on the part of Dalits, their share of illiterates was almost double that of the "general" category having 16 per cent illiterates. However, Dalits were no longer lagging behind OBCs on this count. Among the villages, Senapur witnessed the steepest fall in illiteracy among Dalits, amounting to 10 percentage points, while Pandari made progress on this account by eight percentage points. These two villages made relatively better progress on OBC literacy too.

On the other end of the educational status scale, the share of graduates (including postgraduates) has leapfrogged by an average of seven percentage points from 12.6 per cent to 19.5 per cent for the non-SC–OBC–ST group in a seven-year period. Comparable increase for Dalits was around two percentage points, from 3.5 per cent to 5.7 per cent, and for OBCs, approximately three percentage points, from 3 per cent to 6.2 per cent. The highest quantum of increase in the share of graduates among Dalits was witnessed in Senapur, followed by Gohanakallan. The percentage of OBC graduates grew relatively faster in Gohanakallan and Pandari. Senapur also recorded more rapid growth in the share of graduates belonging to the non-SC–ST–OBC group.

The pattern of educational status also appeared to be related to the landownership structure. The share of illiterates was estimated to be highest among landless households at 36.6 per cent, closely followed by households with a marginal landholding at 34 per cent and 24 per cent for small and above categories of landholdings in 2013. However, illiteracy was highest in the landless category, but the same group of people recorded the highest fall in the share of illiterates during the reference period by four percentage points. The comparable figure for the marginal category was 3.9 percentage points, while no headway was made by larger landownership categories on this account. Two villages, Senapur and Pandari, performed better than the other two in reducing illiteracy among members of landless and marginal households. In comparison, all four villages stagnated on the literacy front among members of small and larger landholdings.

Not just literacy levels but also access to higher education was related to landownership. For instance, the share of graduates (including postgraduates) among members of landless households was only 2.1 per cent in 2013 compared to 4.4 per cent of the marginal category and 9.6 per cent of the small and above category. During the reference period, the inequality among all three classes further widened as the lowest landholding category improved its share by one percentage point against 3.5 per cent of the middle one and 4.5 percentage points of the largest landholding category.

Rising Inequalities

Policymakers around the world continue to grapple with the growth vs inequality dilemma. The dominant view in the debate has consistently advocated the "enlargement of the pie" rather than seeking a larger share of the pie. This view pushed for growth as a priority and everything else could be taken up later. In fact, growth was considered the panacea for all the problems. The growth view seemed to take it as the main fruit of economic processes and inequality as only an inevitable side effect that should be borne for the larger good. Some studies have gone to the extent of claiming that inequality was positively related to socioeconomic indicators. The grow first and distribute later premise has dominated the policies worldwide, including India in the last three decades (Chatterjee, 2021; Rao, 2022; Ghose, 2023).

However, the other strand in the debate contested claims that inequality necessarily accompanied growth. The historical and sectoral analysis formed the basis of the argument that the characteristics of economic growth differed across nations and inclusive or exclusive growth resulted from state policies, among other factors. The broad-based growth galvanised sectors of the economy that provided livelihood to a larger number of people. Progressive state policies on social development and democratisation also reduced disparities between the rich and the poor (Chatterjee, 2021; Rao, 2022; Ghose, 2023).

Nobody disputes the fact of widening inequalities between the rich and the poor in the last three decades, often measured by per capita income, consumer expenditure, and wealth ownership. For instance,

> the average annual real per adult income growth in India accelerated from 1.7 per cent during 1951–80 to 3.3 per cent during 1980–2015. However, for the bottom 50 per cent income group, it decelerated from 2.2 per cent to 1.9 per cent over the same period, despite an acceleration from 1.2 per cent to 5.1 per cent for the top 10 per cent income group and from 0.2 per cent to 6.6 per cent for the top 1 per cent income group.
>
> (Rao, 2022, p. 29)

This data reveals a process of reallocation of incomes from the poor to the rich at the heart of increasing inequalities during the period of neoliberal globalisation.

This came as a reversal of the earlier trend when the redistribution of incomes favoured the poor. The share of the top 10 per cent earners came down by six percentage points to 31 per cent, and that of the bottom 50 per cent went up by three percentage points to 24 per cent during the first three decades after independence. By 2015, not only were the gains made until the 1980s reversed, but the top 10 per cent had surpassed their position of the 1950s to usurp 56 per cent of the share in income while the share of the bottom 50 per cent slumped to a historic low of 15 per cent (Chancel and Piketty, 2019 cited in Sahasranaman and Kumar, 2022).

The rising inequality has often been justified by citing the role of growth in poverty reduction. It has been underlined that the rate of poverty reduction was faster in the latter period than in the earlier phase. This view admits that the poor have been experiencing slower growth in their income than the rich, resulting in widening inequalities, but the poor shouldn't complain about it since they too were economically better off than before. Recent research has shown that the period of all the groups growing, albeit at a differential pace, barely lasted two decades. Since the turn of the millennium, the poor have actually been experiencing negative income growth, a trend that was accentuated during the last decade. All this while, the rich maintained steady income growth (Saharsanaman and Kumar, 2022). For instance, the latest estimates suggested that farmers' incomes declined by 1.5 per cent annually during 2016–2017 and 2020–2021. Casual workers too experienced negative income growth between 2017 and 2022 (Himanshu, 2023).

The growing inequality has proved to be deadlier in rural areas. While the bottom 50 per cent slightly increased their share in income and the top 50 per cent witnessed a corresponding loss in urban areas between 2014 and 2019, the middle deciles gained a share at the expense of the top and bottom deciles in rural India. This might have further worsened the condition of the lower segments of rural areas who were already surviving on an impoverished income. Data indicated that the income share of the bottom layers of rural areas pulled down all-India shares of bottom deciles to the extent that falling shares of bottom segments appeared primarily a rural phenomenon. The bottom deciles were not anonymous. They were essentially composed of small and marginal farmers and agricultural workers. SC and ST were over-represented in these categories and under-represented in higher deciles (Saharsanaman and Kumar, 2022).

Contrary to popular belief, rural societies have seen widening inequalities in contemporary times, over and above pre-existing inequalities. While the ownership of physical and financial assets, particularly land, was considered to be a major source of inequality in earlier periods (Rawal, 2013), with the declining share of agriculture in rural incomes and employment, non-agricultural incomes have added a new inequality-generating source. However, while the relative level of the total income of an agricultural household continued to have a direct relationship with landownership, the share of non-farm income in total income was estimated to be inversely related to landownership (Bathla and Kumar, 2019). Rising income inequalities posed a serious challenge to public policy as access to quality education emerged among the leading determinants besides caste, gender, nature of the job, and work experience (Mehta and Shree, 2017). The crucial facet of contemporary rural inequality was that new layers of inequalities were superimposed on pre-existing ones, instead of replacing them, particularly since the process was mediated by caste and gender.

Income-based estimates have been trending for some time now, but there existed a number of wealth-based inequality estimates. These estimates provided a ready reckoner of household inequalities on the basis of relatively easily collectable data. These estimates have been found more accurate in depicting relative rather than absolute economic status. For instance, recent estimates underscored a steep rise in wealth inequality in the new millennium. This phenomenon could be easily discerned from the falling share of rural wealth at a much faster rate than its share in population, and wealth being concentrated in urban centres. The widespread agrarian crisis and falling share of wages in total output have contributed immensely to the widening inequalities between rural and urban areas and groups of people. The gains from economic growth over the last two decades have largely been cornered as profits by capitalists, while workers' wages did not experience a commensurate increase. The dynamics of wealth concentration ride on the back of pre-existing relations, reflecting disparities among social groups. The shrinking of the asset owner group and the adverse effect of inequality on social outcomes, including access to education and health, have the potential to upturn the economic growth cart (Anand and Thampi, 2016; Chatterjee, 2021).

One of the most powerful theories of growing inequality has been popularly known as R > G, i.e., when the return on capital is greater than rate of economic growth, the reallocation is favourable to capital at the expense of the share of wages in the value of output, producing ever-increasing inequalities (Piketty, 2014). An equally forceful explanation focused on identifying the leading sector of the economy to determine the exclusive or inclusive nature of growth. In this formulation, a divorce between growth and development was the most common symptom of exclusive growth that bypasses the poor. The other consequence of exclusive growth was the rapid sidelining of unskilled and semi-skilled labourers in terms of employment generation quality and quantity. Exclusive growth resulted in the concentration of income in the wealthy class, with the lion's share of incremental income accruing to the richest one-tenth of the population (Ghose, 2023).

As against the hopes of mainstream growth theories, the sectoral balance was rarely maintained in a growing economy due to the fact that growth doesn't trigger in all sectors concurrently. Depending on technological innovations, policy impact, etc., it might originate in some sectors before spreading to other sectors via demand and supply linkages. The capital or labour-intensive nature of the sectors where growth begins was crucial in determining the distribution of income and impulses sent to other sectors via influence exerted on the composition of output and demand for factors

(Ghose, 2023). Further explaining the dynamics of exclusive growth, Ghose (2023, p. 38) maintained that:

> if the lead sectors happen to be those that produce low-priority goods and services, then beneficiaries of growth are the holders of capital and skilled labour, that is already rich. The incremental income then accrues mainly to the rich and generates new demand for low-priority goods and services. Increased production of these goods and services in response to the increased demand means increased employment of capital and skilled labour – the factors held by the rich. The growth process is then driven by the interaction between growing income inequality and shifting production structure towards low-priority goods and services, which means skill bias in employment growth and progressive exclusion of the low-skilled from employment. The economy experiences exclusive growth.

The nature of growth has clear consequences for rural transformation. If the sector leading growth in an economy produces high-priority goods and services, then its benefits would accrue to a larger population due to a completely different demand for factors and semi-skilled and unskilled workers participating in the growth story. When this social formation moves to a high-productivity one based on the workers' skill enhancement, the rural transformation would be qualitatively different (Ghose, 2023). Essentially, it was this notion of rural transformation that we started with, which was a broad-based, employment-generating momentum that leads the entire population to a better life in terms of social development, equity, and freedom from primordial privileges and handicaps. While most of the works cited above have considered a macro picture, the data in this chapter looks at the phenomenon from the other end. It attempts to read the nature of transformation from its impact on the rural population.

Inequality in Ownership of Assets

Ownership of durable households indicates economic progress or lack thereof. More than that, it also reveals inter-group inequalities. The composition of durable assets also shows cultural changes under the impact of market penetration. All these facets essentially indicate the transformation of society.

Assessment of durable household assets is easier said than done. Both the methods applied in estimating the asset base of a household have their lacuna. The factor analysis method has been criticised for ignoring the values of the assets and differentiating them on that basis. At the same time, the estimation of value method was found wanting on account of reporting bias and inaccuracies in the depreciation of different classes of assets.

Table 3.7 shows data on durable assets collected in the 2013 and 2020 surveys. The average value of durable household assets was Rs 21485 in

Table 3.7 Different Social Groups by Relative Distribution of Household Durable Assets, 2013 and 2020

Village/Social Groups	*SCs*		*STs*		*OBCs*		*General*		*Total*	
	2013	2020	2013	2020	2013	2020	2013	2020	2013	2020
Senapur	0.73	0.53	0.37	0.56	0.94	0.61	2.16	3.63	1.00	1.00
Seemli	0.67	0.80	0.15	0.23	1.11	1.15	–	–	1.00	1.00
Pandari	0.48	0.61	0.56	0.24	0.84	0.81	3.31	2.97	1.00	1.00
Gohanakallan	0.71	0.57	0.37	0.05	1.06	1.13	1.64	1.80	1.00	1.00
Total	0.65	0.60	0.33	0.25	1.09	1.04	2.10	2.53	1.00	1.00

Source: UP Rural Transformation Survey (2013, 2020).

Senapur in 2013. This figure was Rs 15690 for SCs, Rs 20300 for OBCs, and Rs 46438 for non-SC–OBC–ST households. The average value of household assets was highest at Rs 41270 in Seemli among all the villages. The monetary value of assets held by SC households was Rs 27710, while OBC households had assets worth Rs 45657 in 2013.

Pandari, the village in Bundelkhand, had the lowest average value of durable assets. Among the social groups, non-SC–ST–OBC category households owned Rs 69015 on average, while OBCs and SCs held assets valued at Rs 17434 and Rs 10030, respectively. This village had the lowest average value of investments and the highest inequality among social groups in the ownership of durable assets. Each household of Gohanakallan possessed durable assets worth Rs 37955. The comparable amount was Rs 62217 for non-SC–ST–OBC households, followed by Rs 40351 for OBCs and Rs 27710 for SCs.

However, comparing two absolute numbers seven years apart is not methodologically advisable. But the objective was not to estimate the total value of assets but to study their distribution. More than the absolute value of the durable households, the relative position of social groups is more revealing. The relative status value of SC, OBC, and "general" category households in Senapur was estimated to be 0.73, 0.94, and 2.16, respectively. This meant that general category households possessed 216 per cent of the average value of durable assets, and SC households held only 73 per cent. OBCs were closer to the average at 94 per cent.

The distribution of durable assets was relatively less skewed in Seemli, with SC and OBC households having values of 0.67 and 1.11, respectively. Pandari was an unequal village regarding the possession of durable assets. The relative values of SC, OBC, and general categories were 0.48, 0.84, and 3.31, respectively. Comparable values for Gohanakallan were 0.71, 1.06, and 1.64. The distribution of assets appeared to be relatively least skewed in the last village.

Inequality in rural societies was on the rise. The data on the relative distribution of durable household assets revealed that, on average, the value of assets held by a non-SC–OBC–ST household was more than four times higher than the value of an SC household and two and a half times that of an OBC household. This gap was three times in the case of SCs and two times in the case of OBCs in 2013.

In Senapur, the relative value of SCs has gone further down by 0.20 percentage points during the reference period, OBCs went down by over 0.30 percentage points, while non-SC–ST–OBC strengthened their relative status from 2.16 in 2013 to 3.63 in 2020. SCs of Seemli slightly improved their relative value of durable household assets, and OBCs marginally went down. In Pandari, SCs slightly improved the score value based on ownership of the value of durable household assets. OBCs remained stagnant, and non-OBC–SC–ST barely came down but still owned five times the average value of assets owned by SC households. In Gohanakallan, OBC and non-OBC–SC–ST marginally improved their status at the cost of SCs, further widening inequalities between the groups.

The relative distribution of financial assets among social groups in 2013 and 2020 indicated that SCs in Senapur had 63 per cent of the average financial asset value of all the village households, compared to 93 per cent of OBCs and 258 per cent of the non-SC–ST–OBC group. This means that a general category household's average financial asset holding was more than four times that of SCs (Table 3.8).

In Seemli, distribution appeared relatively less unequal. SCs had 82 per cent of the average, while OBCs had 110 per cent. OBCs of Pandari were estimated to be holding twice the financial assets held by SCs, while non-OBC–SC–ST groups owned twice that of OBCs and four times that of SCs. In Gohanakallan, the distribution was 55 per cent, 107 per cent, and 214 per cent of the average for SC, OBC, and General categories, respectively, in 2013.

In 2020, compared to 2013, the distribution of financial assets among social groups had changed significantly, further widening existing colossal inequality. In Senapur, SCs had further decreased by 20 percentage points vis-à-vis the average, while the non-SC–ST–OBC group had moved by over 140 percentage points, broadening the difference between both groups to ten times in 2020 from four times in 2013. Seemli, too, witnessed a widening of inequality during the reference period, accompanied by a fall in the relative share of SCs by nine percentage points and a surge in the relative share of OBCs by seven percentage points.

Pandari and Gohanakallan were the two villages where the relative share of SCs had slightly increased in 2020 compared to 2013. In Pandari, this increase was nine percentage points, while Gohanakallan had seen a surge in the relative share of SCs by five percentage points. Since these were close shares, an increase in one group's relative position would be compensated by a fall of another group. In Pandari, this was a non-OBC–SC–ST group whose close share fell by 38 percentage points to 180 per cent of the average value of all households. In Gohanakallan, both OBC and General groups slightly came down to make way for a marginal rise in the relative share of SCs.

Plough is one agricultural asset that has become out of use in all four villages. Bullock is also on its way to becoming extinct in the emerging rural scene. The survey data revealed that machine and power-propelled technology had replaced animal and manual labour-based technology.

However, it was surprising that the number of tractors and trolleys has come down in two villages – Senapur and Pandari, while Seemli has seen substantial growth in the number of tractors and trolleys. Gohanakallan also witnessed a marginal increase. The number of power tillers was also increasing in all the villages. But the number of threshing machines was showing a declining trend. The number of pump sets and sprayers was also decreasing in some villages while witnessing a marginal increase in others.

Table 3.8 Different Social Groups by Relative Distribution of Financial Assets, 2013 and 2020

Village/Social Groups	*SCs*		*STs*		*OBCs*		*General*		*Total*	
	2013	2020	2013	2020	2013	2020	2013	2020	2013	2020
Senapur	0.63	0.42	0.12	0.09	0.93	0.66	2.58	3.98	1.00	1.00
Seemli	0.82	0.73	0.14	0.05	1.10	1.17	–	–	1.00	1.00
Pandari	0.55	0.64	0.22	0.15	1.11	1.13	2.18	1.80	1.00	1.00
Gohanakallan	0.55	0.60	0.21	0.04	1.07	1.01	2.14	2.23	1.00	1.00
Total	0.64	0.56	0.17	0.07	1.13	1.04	1.99	2.68	1.00	1.00

Source: UP Rural Transformation Survey (2013, 2020).

This perplexing scenario needed to be understood with the help of qualitative data. At least two possible reasons could emerge. The state's rural areas have been witnessing the emergence of an agricultural implements rental market. With the subdivision of landholdings and increasing fixed cost of assets, more and more farmers preferred renting them instead of buying them. A connected explanation could be that, with decreasing profitability, it was fast becoming unviable for farmers to possess agricultural implements that needed hefty investments. Given the de-prioritisation of agriculture in rural areas, it was also possible that keeping and maintaining expensive assets was considered unnecessary.

Concluding Remarks

All the studied villages and the country have been undergoing a demographic transition. The change is fast, as reflected in data from two points seven years apart. The changing age structure of the rural population and the expansion of education were two unmistakable patterns across the villages. But the latter was not keeping pace with the former. The share of the population that attained education up to the tenth standard and above, a threshold that would develop the potential of getting into better jobs and upward mobility (Deshpande and Palshikar, 2008), was less than one-fifth in the first survey; it barely crossed the one-fourth mark on the second occasion.

The country has not been able to reap the "demographic dividends" due to the decoupling of education and demographic initiatives. While many countries in the world have been facing the problem of ageing, India and its villages have witnessed a rise in the youth population's share. But the presence of youth in itself could be counterproductive if a comprehensive framework of social development policies and programmes was not in place. Some skill training initiatives have been attempted, but the utilitarian perspective has done more harm than good. The economy-focused policy orientation prepared youth for the industry, not the other way around. It has consequences for the rural transformation of the Indian variety.

Rural India was closely aligned with the rest of the country. But its compulsion to be a junior partner of the urban was conditioned by the continued relevance of the pre-existing structures. Integration with the urban was reflected in dramatic changes in the rural occupational profile, but paradoxically, non-agricultural occupations were not free from the influence of landownership patterns. In other words, the legacy of the land-gender-caste equation continued to hold back vertical mobility and allowed only horizontal mobility.

The sources of widening inequality might be income from sectors other than agriculture, but it has strengthened the divide along caste in close interface with the land. Resourceful people have taken better advantage of the opportunities in the non-agricultural sector. Land and caste were the two major determinants of access to education, ensuring that people moved from the old occupational pattern to the new opportunity structure without disturbing the relative status between groups. This was a colossal failure of the state, as education continued to be intertwined with caste, land, and gender.

The withdrawal of women from the workforce has been a phenomenon of gigantic proportions – the most widely held view attributes this change to economic mobility. While not rejecting this hypothesis, one must caution against an aggregated approach. Historically, women's involvement in labour outside the home has indeed been conditioned by the household's social and economic standing, and it changed in tandem with the mobility experienced by the family.

But this was just one of the explanations and covered only specific segments. But women agricultural labour did not simply withdraw because of upward mobility; farm mechanisation displaced them. Men, too, were displaced, but a group of them found opportunities to work with machines. Similarly, low-skilled women workers did not find openings in the fast-changing technological environment of non-agricultural enterprises that dislodged old-style enterprises that employed women. Women workers also emerged as net losers in newly set up supply chains connecting rural and urban areas.

The transformation occurring in rural areas was inhospitable to low-skilled workers, a larger segment constituted by women. Male workers, too, were displaced, but either they acquired new skills or migrated to far-flung places in search of jobs. Women were excluded from both these opportunities. The character of the transformation was not enabling for those left out. It simply displaced those who did not possess the desired skill sets. But it did not facilitate their acquiring the required expertise. Open unemployment allowed the emerging economy to get away with this responsibility.

However, the technological changes alone were not determinant for the occupational outcomes of the population. The massive "withdrawal" of women from "work" was just another symptom of the spatial dislocation of production. The rural was losing out to the urban in retaining its productive capacities. What was referred to as the falling contribution of agriculture to GDP at an aggregate level translated into the loss of livelihood opportunities in everyday life. As discussed in other chapters, economic activities in rural regions stagnated in some areas and have come to a grinding halt in others. The crisis of complete separation between the economy and the people has been disproportionately distributed among people based on their social location.

The dynamic of transformation was also partial in that it changed economic relations rapidly, at times only modifying and, on other occasions, strengthening the pre-existing social relationships. Both gender and caste continue to influence outcomes in the newer economy. Often, the state was expected to play an active role in addressing these anomalies by supporting the marginalised groups. The state's role was rather limited to ensuring working people's survival, even at a subhuman level. The pathways of transformation in the West and India couldn't be more different.

Integrating the village economy with the macroeconomy has acted as a double-edged sword. While it has gradually improved the absolute socio-economic conditions of working people, it has widened relative gaps between the groups.

The new opportunities reduced the dependence of the poor on the agrarian structure but exposed them to heightened vulnerabilities in cities. Despite spatial shifts, they continued to form the bottom of the heap.

The rural transformation due to integration with the urban did not accompany a radical disruption of the foregoing order. The pre-existing economy provided a stepping stone for entry into a new economy.

4 Gender, Capital, and Occupational Change

The occupational pattern of a population has been considered a vital indicator in transformation studies, more so in studies on rural transformation. For an extended period, occupational shifts were discussed within the framework of agriculture to non-agriculture. However, contemporary works identify complexities beyond this binary. The assumption of a predetermined and linear progression has been abandoned in favour of a more nuanced approach that recognises the multiplicity and simultaneity of occupations.

Occupational changes could be probed as occupational shifts in the working-age population, inter-locational occupational shifts, and most interestingly, intra- and intergenerational occupational mobility. Since South Asia, in general, and India, in particular, have a long history of caste and occupation linkage, occupational shifts co-constitute transformation. But evidence on occupational change was quite contradictory and plagued by definitional variance. One of the most contentious issues of caste as the axis of upward or downward mobility divided opinions in social science works on the subject (Iversen et al., 2017). For instance, Kumar et al. (2002) argued that class origin markedly determined class destinations, while caste was not the most relevant category in the analysis of mobility. This was countered by Deshpande and Palshikar (2008), who argued that caste might not matter in downward mobility, but it certainly was an element in upward mobility.

From the point of view of transformation, the study made a pertinent point that different groups might have diverse trajectories of mobility. They attribute occupational change among Dalits to affirmative actions of the State and upwardly mobile peasant castes to their resource base. However, the relationship between caste and mobility was rather complex. While urban spaces provided opportunities for upward mobility irrespective of caste, moderating the deficits created by location at the lower stratum of caste, the study did not find evidence of caste's role in downward mobility, and it concluded that caste mattered in cases of upward mobility. But more than that, the authors questioned the uncritical superimposition of caste on any economic development, including downward mobility caused by the dynamics of capitalism (Deshpande and Palshikar, 2008).

Other investigations have also presented a mosaic of evidence on mobility (Jarvis and Song, 2017). For instance, some contemporary studies have

DOI: 10.4324/9781003588245-4

found evidence of "convergence in mobility rates of SC/STs and non-SC/STs to non-SC/ST levels in both education attainment and wages" (Hnatkovska, Lahiri, and Paul, 2013, p. 468) and "strong intergenerational stickiness in both educational achievement and occupational distribution among SCs and STs" (Majumdar, 2010, p. 463). Further, studies on mobility among Dalits and Adivasis found increasing consumption expenditures and a sense of well-being among both marginalised groups, but other investigations underlined the persisting practice of untouchability and discrimination in the job market (Shah et al., 2006; Soundararajan et al., 2023; Thorat and Dubey, 2012; Iversen et al., 2017).

However, the caste question made a comeback in other studies in the form of points of origin debate. The starting point of mobility might not exclusively consist of caste location; it was an essential ingredient along with land in rural areas, parental income in urban spaces, and gender, a relevant category across locales. Since mobility in India is rather slow, it didn't successfully break the glass ceiling imposed by descent. Some studies, for instance, Deshpande and Palshikar cited above, testified to a seemingly equivalent pace of mobility among groups, but that was more of an admission of persisting inequalities and resilience of sources of inequalities (Iversen et al., 2017).

Like transformation, occupational mobility wasn't a one-way street for all. Often, the Indian situation was marked by downward mobility caused by medical and income crises. The unevenness of mobility for groups could be seen in the faster pace for some, slow for many others, and falling back for a few. This wasn't the case only among social groups, but more so among genders. Frequently, women had to resort to traditional occupations during men's transition to modern callings (Iversen et al., 2017). The fragility of mobility could also be seen in many other instances (Ahmed and Nauriyal, 2023).

These studies essentially posed questions of occupational mobility and mobility in general that every study of transformation would have to face. These related to relationships with caste, gender, and class; variance in trajectories of mobility; implications of location; the pace of mobility; and risks of downward mobility, and they reminded that these interplays were taking place in the context of capitalism. But these studies miss one crucial point. While their complaint about the superimposition of caste on economic developments might have some merit, they elaborated the data, often going back generations, describing the diverse trajectory undertaken by different caste groups. For instance, Deshpande and Palshikar (2008, p. 65) underlined that upper castes consolidated their position during the course of occupational mobility over four generations, the Marathas moved from "limited advantage" to "spectacular advantage", OBCs remained "prisoners" of the "lower-middle" occupations, and Dalits only experienced "superficial" upward mobility. Moreover, all these developments occurred in a modern city like Pune with a booming market economy. Despite speaking evidence, the authors did not attempt to theorise the linkage between caste and capitalism.

A number of studies in this genre have shied away from questions on how capitalism and caste have modified each other to stay relevant.

Further, mobility studies frequently failed to situate changes in an emerging social milieu marked by structural transformation of the economy, technological innovations, employment patterns, skill endowments, etc. However, some of the studies have remarkably contextualised intergenerational mobility in prevailing socio-political contexts; the political economy of the employment and spatial patterns often skipped the attention. Mobility should be understood with reference to macroeconomic changes. If a social formation was transforming, all its constituents would likely experience a change. It might result in horizontal mobility if the group's position did not change vis-à-vis other groups. Only when a group's relative positions against other groups changed, vertical mobility in the forms of upward or downward mobility could be spotted.

This proposition also connects the dots between capitalism-caste congruence and the starting point argument. While the latter contention essentially pointed out that the relative gap between the two groups at the starting point might outweigh the entire trajectory of mobility, focusing on capital's relationship with primordial structures, particularly caste, would help understand the phenomenon. For instance, if the comparative distance on the occupational scale between UCs and Dalits exuberated in the course of mobility, it could indicate a mutually supportive relationship between caste and capitalism.

Gender constitutes another axis around which occupational mobility or lack thereof revolves. The most common form in which gender interferes with occupational mobility was gender-segregated occupational patterns. Studies refuted educational achievements as the factor for different labour market outcomes, pointing out the narrowing gap in human capabilities between women and men amidst the persisting wage gap and employment segregation. These probes raised questions on the nature of growth that has failed to increase women's participation in paid employment. On the contrary, India has been witnessing the withdrawal of women into household activities at the expense of female work participation in the labour market. These investigations also underlined that the enduring wage gap between men and women could not be understood by productivity factors but partly by gendered occupational segregation and discrimination in the labour market (Das and Saha, 2022).

The gendered discrimination in the labour market closely interfaced with societal prejudices about the lack of ability among women to perform certain tasks and roles. In fact, stereotypes, entry barriers, and under-representation formed a vicious cycle, strengthening gender-based labour regimes, consequently laying the ground for further discrimination and occupational immobility (Kulkarni and Hatekar, 2013).

However, gender biases were not the only hindrance; the withdrawal of women from the workforce testified to a thick congruence between caste, gender, and class. This was particularly apparent in the case of lower castes, where women have traditionally been working outside the home. As these households ascended on an economic scale, they tended to imitate upper-caste practices of confining women to homes and began to withdraw them from paid work. This phenomenon runs the risk of being misconstrued as increasing the comfort of working-class women. After all, who wants to engage in back-breaking work? That might be true to an extent, but it came at the cost of freedom and status within the family that these women had been enjoying as economic contributors. Cohn (1961, p. 1053) had aptly noted in Senapur:

> A Thakur woman is directly under the control of the older women of the house, while a Chamar woman is out working and mixing with men. The Chamar woman has a view of herself as a person apart from her husband. She is a full and active economic contributor to her household...In most Chamar households, the economic roles of husbands and wives are interchangeable.

The withdrawal of women from the workforce has often been misapprehended as an economic phenomenon since it coincides with economic mobility. Women were not simply withdrawn from economic activities outside the home but also brought under the control of men. It was essentially a socio-political process that demonstrated the dominance of the upper castes' cultural ethos over the entire society and the lower castes' eagerness to raise social status by resorting to keeping women under check as soon as they acquired the wherewithal to make do without women's earnings. Against mainstream assumptions, occupational mobility was not a linear and unidirectional process. Depending on its interaction with social relations, it might take various twists and turns. It did not happen in a vacuum but was shaped by social structure.

Besides missing the interface of caste, gender, and class, many of the works have considered only the occupation of the head of the household or male/female members in the analysis of intergenerational mobility. The estimates of mobility, whether intergenerational or intragenerational, only based on limited coverage of household members could be misleading. Data based only on primary occupation would also fail to yield accurate analysis in the contemporary scene marked by multiple occupations.

Considering the fallacies of the existing works on the subject, this study collected and analysed primary and secondary occupation data on two points to estimate intragenerational mobility. All the working-age household members were covered in this estimation. More importantly, combinations of primary and secondary occupations in 2013 and 2020 have been analysed to understand the changing livelihood strategies of rural workers. In the course of the analysis, gender emerged as the most crucial explanatory factor in occupational change and has been given detailed treatment in the chapter.

Occupational Change in Studied Villages

Cultivation was reported by 26 per cent of household members of the working-age group in 2013 as their primary occupation in Senapur (Table 4.1). Another 31.5 per cent reported it as their secondary occupation. Non-agricultural labour (28.9 per cent) was the most significant primary occupational category in 2013, followed by animal farming (20.7 per cent), agricultural labour (11.8 per cent), salaried jobs (7.6 per cent), and trading (4.7 per cent). Non-agricultural labour came only in 11.7 per cent of persons as a secondary occupation. Agricultural labour, too, had a relatively smaller report at 9.9 per cent. But animal farming was reported as the secondary occupation of close to 46.5 per cent of working-age persons.

Seemli's occupational pattern was not entirely different from Senapur's. However, the former was located in the western part of the state, while the latter was in eastern UP. About 30.4 per cent of the working-age population in this village reported animal farming as their primary occupation, followed by 23.3 per cent for cultivation, 21.5 per cent for non-agricultural labour, and 16.5 per cent for salaried jobs. Agricultural labour was reported by less than 5 per cent of the working-age population. Alongside, 51 per cent of persons were engaged in animal farming as their secondary occupation, followed by 27 per cent in cultivation, 13.7 per cent in agricultural labour, and 7.6 per cent in non-agricultural work.

Non-agricultural labour as the primary occupation was reported by 30.6 per cent of persons of working age in Pandari, followed by 29 per cent in animal farming and 25.6 per cent in cultivation. Fewer persons reported salaried jobs (5.6 per cent) and trading (2.6 per cent). Cultivation appeared to have been reported as a secondary occupation by 46.1 per cent, followed by animal farming by 28.2 per cent, agricultural labour by 15.6 per cent, and non-agricultural labour by 9.4 per cent.

Gohanakallan is another village that followed the prevailing pattern of non-agricultural labour, reported as the primary occupation by the most significant number of persons in 2013. Animal farming (27 per cent) came out as the second-leading occupation, followed by cultivation (22 per cent), salaried jobs (11.5 per cent), and trading (6.9 per cent). In this village too, cultivation was reported by a more significant number of persons (51.5 per cent) as a secondary occupation. In contrast, non-agricultural labour was registered as a secondary occupation by a fraction (3.8 per cent) of persons. Animal farming (35.3 per cent) too came as a secondary occupation for a sizable number.

Senapur was the only village where a visible drop in working-age persons reporting cultivation as their primary occupation was noticed during the reference period. In Pandari, it was also almost at the same level in 2020 as in 2013. Seemli recorded an increase of three percentage points, while Gohanakallan saw a jump of five percentage points in reporting cultivation as a primary occupation. While taking all four villages together, the share

Table 4.1 Population Distribution by Primary Occupation: 2013 and 2020

Occupation Category	*Senapur*		*Seemli*		*Pandari*		*Gohanakallan*		*Total*	
	2013	2020	2013	2020	2013	2020	2013	2020	2013	2020
Cultivation	26.17	19.33	23.37	26.27	25.64	26.14	22.05	27.07	24.07	24.39
Animal Farming	20.73	2.12	30.41	5.49	28.85	1.33	26.97	3.34	26.64	2.98
Agricultural Labour	11.8	4.49	4.98	2.35	6.25	3.41	4	1.98	6.51	3.1
Casual Non-Agri Labour	28.9	37.91	21.48	36.08	30.61	52.65	28.21	33.37	27.52	39.11
Handicraft	0.15	4.61	0	3.33	0.32	0.76	0.31	2.47	0.21	2.94
Cottage Industry	0	1.5	0.17	2.75	0.16	0.38	0.1	0.87	0.11	1.32
Trading	4.69	5.49	3.09	4.51	2.56	3.03	6.87	5.93	4.64	4.95
Salaried Job	7.56	24.56	16.49	19.22	5.61	12.31	11.49	24.97	10.31	21.22
Total	100	100	100	100	100	100	100	100	100	100

Source: UP Rural Transformation Survey (2013, 2020).

of cultivators primarily remained almost stagnant in both surveys. A quick jump in the percentage of cultivation in two villages might reflect the Covid-2019 lockdown-led return of migrants who had nothing else to fall back upon other than tiny family farms (Table 4.2).

Another noticeable aspect is a sharp increase in the share of cultivation in the workforce as a secondary occupation. In two villages, Senapur and Seemli, this increase was in the range of 16–17 percentage points, indicating that workers were not quitting cultivation but diversifying while keeping a toehold in it. In Pandari, this jump was eight percentage points. At the same time, Gohanakallan had already reached a low share of cultivation as a primary occupation but very high reporting of the sector as a secondary occupation in 2013.

But overall, cultivation kept its share of the workforce occupied during the reference period. Possibly, its share had already fallen significantly before 2013, and a further fall would require a break in the economic or technological sphere.

While the share of cultivation among workers had already seen a dramatic transformation in the decade leading to the first survey in 2013, animal farming witnessed equally upsetting changes during the reference period. Its share as a primary occupation among the workers fell sharply during the seven years between the surveys. This fall was 18 percentage points in Senapur, 25 percentage points in Seemli, 27 percentage points in Pandari, and 23 percentage points in Gohanakallan. After this fall, workers primarily engaged in animal farming decreased to a new low of 1.3 per cent in Pandari, 2.1 per cent in Senapur, 3.3 per cent in Gohanakallan and 5.5 per cent in Seemli.

Simultaneously, a considerable fall in reporting of animal farming as a secondary occupation across villages was observed. It saw a drop of 37 percentage points in Senapur, 10 percentage points in Seemli, and three percentage points in Pandari. In Gohanakallan, the share of workers engaged in animal farming went up by 2.5 percentage points. Overall, animal farming has witnessed a sharp fall in its share of workers during the reference period. Earlier, cultivation and animal farming went hand in hand. While cultivation has been losing its primary occupational status in the contemporary scene, animal farming has almost lost its secondary status.

The large pool of manual labour in rural areas had two groups, closely integrated into each other. The ones who were confined to providing labour to agricultural operations, and others who chose to work outside agriculture too, both in rural and urban areas. The first was generally referred to as farm labour, and the second was casual non-agricultural labour; however, these two groups were not watertight compartments.

Except in Senapur, where nearly 12 per cent of workers reported agricultural labour as their primary occupation in 2013, all other villages already had a meagre share of this category, averaging 6.5 per cent for all four villages taken together. This figure fell to 3.1 per cent in 2020. Senapur experienced a fall of seven percentage points, and the share of agricultural labour

Table 4.2 Population Distribution by Secondary Occupation: 2013 and 2020

Occupation Category	*Senapur*		*Seemli*		*Pandari*		*Gohanakallan*		*Total*	
	2013	2020	2013	2020	2013	2020	2013	2020	2013	2020
Cultivation	31.5	48.7	27.0	43.8	46.1	54.1	51.5	51.2	42.3	50.1
Animal Farming	46.5	9.6	51.2	40.9	28.2	25.0	35.3	37.8	37.9	29.5
Agricultural Labour	9.9	22.1	13.7	8.9	15.6	10.3	8.7	4.6	11.6	10.3
Casual Non-Agri Labour	11.7	15.5	7.6	3.0	9.4	8.2	3.8	4.2	7.6	7.3
Handicraft	-	0.4	–	1.3	–	0.0	–	0.2	–	0.4
Cottage Industry	0.0	1.1	0.0	0.4	0.0	0.3	0.2	0.0	0.1	0.4
Trading	0.0	1.1	0.0	0.4	0.7	0.9	0.5	0.6	0.4	0.7
Salaried Job	0.3	1.5	0.5	1.3	0.0	1.2	0.0	1.4	0.1	1.3
Total	100.0	100.0	100.0	100.0	100.0	100.0	100.0	100.0	100.0	100.0

Source: UP Rural Transformation Survey (2013, 2020).

among all workers fell by approximately two percentage points in each in all other villages.

The steep fall in reporting of agricultural labour as a primary occupation appeared to have swelled the reporting of it as a secondary occupation in Senapur. A segment of workers has dropped agricultural labour as their primary occupation but retained it as their secondary occupation. The reporting of agricultural labour as a secondary occupation has increased by more than ten percentage points in Senapur. In the other three villages, it came down by five percentage points (Seemli and Pandari) and four percentage points (Gohanakallan).

The Combinations of Primary and Secondary Occupations

Table 4.3 presents the distribution of secondary occupations according to the reporting of the primary occupation. Rows depict secondary occupations, and columns depict primary occupation.

In 2013, among all those who reported cultivation as their primary occupation, over 77 per cent reported animal farming as their secondary occupation, and another 15.5 per cent reported non-agricultural labour as their secondary occupation. Around 84 per cent of those who preferred animal farming as their primary occupation opted for cultivation as their secondary occupation. Approximately 11 per cent of workers primarily occupied in animal farming reported agricultural labour as their secondary occupation. Less than 5 per cent of such persons were engaged in non-agricultural labour as their secondary occupation. However, their number might be limited, but for those primarily engaged in agricultural labour, close to half of them were in animal farming too, followed by 28 per cent in non-agricultural labour and 24 per cent in cultivation.

All the working-age persons who reported non-agricultural labour as their primary occupation also reported cultivation (56 per cent), agricultural labour (25 per cent), and animal farming (19 per cent) as their secondary occupation.

By 2020, among those who opted for cultivation as their primary occupation, the share of those occupied in animal farming as a secondary occupation had gone down by 24 percentage points. Their share in non-agricultural labour had gone up by 15 percentage points. But at the same time, for those primarily occupied in animal farming, their participation in cultivation as a secondary occupation increased at the cost of involvement with agricultural labour. A larger share of farming labourers involved themselves in cultivation in 2020 compared to 2013, while their participation in animal farming appeared to have gone down. Similarly, around 85 per cent of all those who chose non-agricultural labour as their primary occupation were involved in cultivation, almost leaving agricultural labour and animal farming, activities in which a sizable number of them were occupied in 2013.

Table 4.3 Distribution of Observations of Secondary Occupation as Per Primary Occupation, 2013 and 2020

Primary → *Secondary* ↓	*Cultivation*		*Animal Farming*		*Agri Labour*		*Non-Agri Labour*		*Handicraft*		*Cottage Industry*		*Trading*		*Salaried Employment*		*Total*	
	2013	2020	2013	2020	2013	2020	2013	2020	2013	2020	2013	2020	2013	2020	2013	2020	2013	2020
Cultivation	0.4	3.3	83.8	87.5	24.4	53.8	55.7	84.6	40.0	94.4	33.3	80.0	67.1	83.7	74.1	80.0	42.3	58.7
Animal Farming	77.3	53.5	0.0	0.0	47.8	15.4	19.3	5.2	40.0	0.0	66.7	20.0	30.1	14.0	25.0	9.5	37.9	21.2
Agri Labour	5.7	6.5	11.4	8.3	0.0	0.0	24.5	7.7	0.0	5.6	0.0	0.0	1.4	0.0	0.9	1.1	11.4	5.7
Casual Non-Agri Labour	15.5	29.8	4.4	4.2	27.8	30.8	0.0	0.0	0.0	0.0	0.0	0.0	1.4	2.3	0.0	6.3	7.7	10.9
Handicraft	0.0	0.5	0.0	0.0	0.0	0.0	0.0	0.3	0.0	0.0	0.0	0.0	0.0	0.0	0.0	0.0	0.0	0.3
Cottage Industry	0.2	0.9	0.0	0.0	0.0	0.0	0.0	0.3	0.0	0.0	0.0	0.0	0.0	0.0	0.0	0.0	0.1	0.4
Trading	0.6	2.8	0.3	0.0	0.0	0.0	0.4	0.3	20.0	0.0	0.0	0.0	0.0	0.0	0.0	2.1	0.4	1.3
Salaried Employment	0.4	2.8	0.0	0.0	0.0	0.0	0.0	1.4	0.0	0.0	0.0	0.0	0.0	0.0	0.0	1.1	0.1	1.6
Total	100.0	100.0	100.0	100.0	100.0	100.0	100.0	100.0	100.0	100.0	100.0	100.0	100.0	100.0	100.0	100.0	100.0	100.0

Source: Rural Transformation Survey (2013, 2020).

Gendered Change

The share of working-age males in Senapur fell around 12 percentage points between 2013 and 2020 for those occupied in cultivation, over 13 per cent for animal farming, and 5.5 per cent for agricultural labour (Tables 4.4). The shifting of working-age males appeared to have taken place towards non-agricultural labour, which experienced growth of seven percentage points, and salaried employment grew by an enormous 16 percentage points.

A similar trend was evident in Seemli too, which falls in another region of the State. In this village, an eight-percentage-point drop in cultivation, a fall of equal magnitude in animal farming, and a 1.5 per cent fall in agricultural labour for males of working age were accompanied by an 11-percentage-point increase in the share of the category of casual non-agricultural labour. Salaried employment was already relatively high in 2013.

Pandari's story is also not significantly different where a four percentage points fall in the share of male persons occupied in cultivation and almost vanishing of categories of animal farming, and agricultural labour resulted in further swelling of the category non-agricultural labour by a leap of 16 percentage points reaching to 58 per cent in 2020. The share of salaried employment among males of working age in the village has also doubled during the reference period.

While the share of casual non-agricultural labour in Gohanakallan remained stagnant for seven years between the surveys, males of working age who had withdrawn from animal farming appeared to have shifted to salaried employment.

The story of women is very different from men's (Tables 4.5). Contrary to the pattern of males withdrawing from cultivation, the share of females of working age occupied in cultivation has seen a massive increase during the seven years between 2013 and 2020 in all the villages. In Gohanakallan, this jump was 27 percentage points, followed by 26 percentage points in Seemli, 13 percentage points in Pandari, and six percentage points in Senapur. Overall, the share of women occupied in cultivation has increased from 19.4 per cent in 2013 to 37 per cent in 2020.

In another surprising observation from the two points data, female participation in animal farming has decreased drastically from 50 per cent in 2013 to 16.4 per cent in 2020, taking all four villages together. This fall was 41 percentage points in Pandari, 37 percentage points in Gohanakallan, 35 percentage points in Seemli, and 23 percentage points in Senapur.

While the share of males in agricultural labour has gone down significantly, female participation in this occupation has increased considerably from 10.4 per cent in 2013 to 17.6 per cent in 2020. This increase was 15 percentage points in Pandari, seven percentage points in Senapur, and five percentage points in Gohanakallan. At the same time, the share of female agricultural labour has remained at 7 per cent in Seemli at both points.

Table 4.4 Occupational Distribution of Male, 2013 and 2020

Occupation Category	*Senapur*		*Seemli*		*Pandari*		*Gohanakallan*		*Total*	
	2013	2020	2013	2020	2013	2020	2013	2020	2013	2020
Cultivation	29.98	18.14	32.23	24.37	28.18	24.89	23.69	24.37	27.89	22.6
Animal Farming	13.67	0.43	9.34	1.37	14.09	0.21	13.94	2.1	13	1.08
Agri Labour	7.19	1.57	3.01	1.59	3.87	0.43	1.39	0.56	3.68	1.03
Casual Non-Agri Labour	34.29	42	29.82	41.23	42.54	58.15	38.33	37.39	36.56	43.68
Handicraft	0.24	4.71	0	2.51	0.55	0.86	0.35	2.66	0.3	2.89
Cottage Industry	0	1.57	0.6	2.96	0.28	0.21	0	0.98	0.18	1.38
Trading	5.76	5.86	3.31	4.78	3.59	2.79	8.01	6.02	5.58	5.09
Salaried Employment	8.87	25.71	21.69	21.18	6.91	12.45	14.29	25.91	12.82	22.25
Total	100	100	100	100	100	100	100	100	100	100

Source: Rural Transformation Survey (2013, 2020).

Table 4.5 Occupational Distribution of Female, 2013 and 2020

Occupation Category	*Senapur*		*Seemli*		*Pandari*		*Gohanakallan*		*Total*	
	2013	2020	2013	2020	2013	2020	2013	2020	2013	2020
Cultivation	21.63	27.45	12.02	38.03	22.82	35.48	19.83	47.37	19.39	36.97
Animal Farming	36.33	13.73	65.38	30.99	51.04	9.68	49.72	12.63	50	16.36
Agri Labour	17.55	24.51	7.69	7.04	10.37	25.81	7.26	12.63	10.46	17.58
Casual Non-Agri Labour	16.73	9.8	6.25	4.23	11.62	11.29	11.73	3.16	11.79	6.97
Handicraft	0	3.92	0	8.45	0	0	0.28	1.05	0.1	3.33
Cottage Industry	0	0.98	0	1.41	0	1.61	0.28	0	0.1	0.91
Trading	3.27	2.94	2.88	2.82	0.41	4.84	4.75	5.26	3.04	3.94
Salaried Employment	4.49	16.67	5.77	7.04	3.73	11.29	6.15	17.89	5.13	13.94
Total	100	100	100	100	100	100	100	100	100	100

Source: Rural Transformation Survey (2013, 2020).

Female participation in casual non-agricultural labour has decreased, which again diverged from the occupational pattern of males. Regarding the average of all four villages, this share dropped from 11.8 per cent in 2013 to 7 per cent in 2020. This drop was eight percentage points in Gohanakallan, seven in Senapur, and two in Seemli.

It is only the category of salaried employment where both males and females have registered an increase during the reference period. The average share of salaried jobs for working-age females has increased from 5.1 per cent in 2013 to 14 per cent in 2020. The magnitude of the increase was 12 percentage points in Senapur, 11 percentage points in Gohanakallan, eight percentage points in Pandari, and one percentage point in Seemli.

It appears that while males were moving to non-agricultural occupations, both casual and salaried, females were filling the space vacated by them in cultivation and agricultural labour.

Along the lines of the overall picture, social group-wise disaggregated data also reveal a drastic overhauling of the occupational pattern in all the studied villages, marked by stability in the category of cultivation, an extreme scaling down of animal farming, and an enlargement of salaried employment. Comparing two points of data, the share of animal farming in the primary occupation of Dalits came down to 2.4 per cent in 2020 from 21.8 per cent in 2013. A fall of a similar magnitude was also estimated for OBC and the "general" category in all four villages.

The category of agricultural labour was already non-existent for the non-OBC–SC–ST group in 2013. The share of Dalit workers primarily occupied in it came down to 2.4 per cent in 2020 from 11.3 per cent in 2013, and that of OBCs remained stagnant at around 3 per cent.

The Dalit workers withdrawing from animal farming and agricultural labour as a primary occupation appeared to have shifted to casual non-agricultural labour, raising its share to 53.7 per cent from 36.5 per cent over seven years. The percentage of OBCs also rose from 22.4 per cent in 2013 to 31.9 per cent in 2020. However, the occupational pattern of the non-OBC–SC–ST group was markedly different from the other two groups in terms of continued low participation in non-agricultural labour.

Most of this group was estimated to be primarily occupied in salaried jobs. OBCs and Dalits have also increased their participation in salaried employment. One could safely say that, keeping most of the workers occupied, casual non-agricultural labour has acquired the same meaning for Dalits as salaried jobs for the "general" category. This data revealed a significant pattern of shifting from an agricultural economy to a non-agricultural economy marked by caste. It was clear that all social groups were diversifying outside agriculture, but their location in the agrarian economy reflected on, if not determined, their destination in the non-agrarian economy. Dalits, mainly from marginal farming and agricultural labour, transited to casual non-farm jobs. At the same time, members of the "general" category owning larger tracts of land primarily shifted to salaried employment.

Alongside dramatic changes in the primary occupation of all social groups, Dalits also experienced a massive revamp of their secondary occupation. The share of cultivation in secondary occupation surpassed the halfway mark to reach 53.6 per cent, while animal farming and agricultural labour witnessed a noteworthy fall. The other two social groups did not experience significant changes in the pattern of secondary occupations except for a fall in the share of animal farming for the non-OBC–SC–ST group.

Concluding Remarks

Against the expectations of many, cultivation continued to engage about a quarter of rural workers as their primary occupation in all four studied villages. There might have been a steep fall in this proportion before the first survey, but its share remained steady during the reference period. Besides, another two-fifths of workers in 2013 and about half in 2020 reported cultivation as their secondary occupation. If we add another one-tenth of the workers engaged in agricultural labour as a secondary occupation, the share of rural workers in the agrarian economy would be well above two-thirds, almost touching the three-fourths mark.

Large data sets often designate these households as agriculture-dependent or farming households. However, one could easily discern that the empirical situation is different. Income estimates have not been provided here, but if occupational choices indicated households' dependence on sources of earnings, agriculture has slipped from its prime position in rural north India.

But despite these massive changes, "post-agrarian rural" is an idea whose time has not yet arrived. Nor could we term it a complete "disjunction of rural and agrarian". Agriculture might not have remained the biggest source of employment in rural India, but about three-fourths of workers had an economic interest in it. More than that, it was a shock absorber. One could see an uptick in the reporting of cultivation in 2020. In all likelihood, it was returnee migrants who came home devastated after the Covid-19 lockdown to survive on cultivation until things normalised.

While prophecies on the demise of cultivation as an employment source could wait for the time being, its closest ally, animal farming, has suffered a jolt in the occupational choices of rural workers during the reference period. While it has almost vanished from the primary occupation, a sizable share of workers has also opted out of it as a secondary occupation. One might argue that close to 30 per cent of workers' involvement in secondary occupational capacity did not remotely signify the downsizing of the animal farming economy, but it essentially slid in workers' priority, possibly due to diminishing returns or availability of better options.

Similarly, the reporting of agricultural labour as a primary occupation has further declined; it continued to be relevant as a secondary occupation for about one-tenth of rural workers in studied villages. However, agricultural labour was never a viable occupation, and workers invariably attempted to combine it with other sources of income like tenant farming, animal husbandry, or non-agricultural labour. Two major developments that have further diminished the prospects of agricultural labour as the primary occupation. First, the technological changes due to the expansion of public and private irrigation infrastructure, leading to increased cropping intensity and mechanisation of several agricultural operations, have shortened period of employment opportunities in a year for manual workers.

Second, the withdrawal of Dalit workers from agricultural labour was primarily a political action rather than an economic decision since their dependence on UCs for livelihood came in the way of opposing caste domination. The data from this study has sufficiently revealed that agricultural labourers primarily belonged to Dalit and other lower caste communities, and working on the farms of dominant castes was not simply an economic activity.

While agrarian occupations tended to lose the centrality they once had, the shifting occupational base of rural workers was evident in the ascendancy of non-agricultural occupations. Casual non-agricultural labour had already become the largest occupational choice of rural workers by the time of the first survey; it further consolidated its position by engaging almost two-fifths of workers in 2020. Remarkably, it was a primary occupation, with a small share of workers opting for it in a secondary status. This role reversal happened not so long ago; otherwise, people tried their hands at off-farm casual jobs during lean agricultural seasons. It was considered a supplementary source of income for those below the middle peasantry classes on the agrarian hierarchy.

From nowhere to becoming predominant, the off-farm jobs tapped resources reeling under the perpetual agrarian crisis over three decades since the 1990s. The hidden unemployment in agriculture was no hidden secret during the post-green revolution mechanisation spree that has been replacing agricultural operations, one after another at a breakneck speed. And consistently falling real income from the agrarian economy made the matter worse.

Something that often escaped the attention of economists was that, alongside diminishing economic prospects, agriculture was fast losing its prestige. Not many people took pride in calling themselves a *Kisan*, frequently perceived as the last resort of the unsuccessful, who failed to carve out a place for themselves in expanding urban jobs that brought home respect and cash.

The booming city economy came as a respite for the rural workers who were already tired of reporting themselves occupied in agriculture but knew that their farms would do equally well without them. The icing on the cake was that jobs in construction, petty manufacturing, and small-scale services offered much-needed cash, something the agricultural economy never supplied in sufficient quantity, even during its heydays. The working of both push and pull factors in tandem ensured that off-farms quickly overshadowed farms in their bastion.

Whereas casual off-farm jobs predominantly absorbed the rural underclass, the privileged classes opted for salaried employment in private and public enterprises. The share of salaried jobs in the occupational choices of rural workers has doubled during the seven years to cross the one-fifth mark. It was not surprising that regular employment was exclusively chosen as a primary occupation, with almost negligible reporting as a secondary occupation. The last indication of a shifting occupational base was seen in the ascendancy of trading in the primary occupational choices of the rural workers.

Overall, non-farm casual and regular jobs and trading were gaining space at the expense of cultivation, animal farming, and agricultural labour that were either receding or remained stagnant, according to the data drawn from all four villages. Having said that, many rural workers were still occupied in the agrarian sector, debunking the possibility of "post-agriculture rural".

The transforming occupational structure of studied villages, focusing on two points, was also reflected in the combinations of primary and secondary occupations reported by workers. A clear shift in the secondary occupational choice from animal farming to casual non-agricultural labour was visible for workers primarily occupied in cultivation. However, animal farming was still cultivators' preferred second option, but its share fell substantially alongside a commensurate rise in off-farm jobs. On the other side, most animal farmers persisted with the choice of cultivation as a second occupation. However, in their case, a slight shift took place in favour of non-farm jobs at the expense of farm labour.

Surprisingly, cultivation has replaced animal farming as the preferred secondary occupation of agricultural labourers. They also appeared to be leaning further towards off-farm jobs while withdrawing significantly from animal farming. All those looking forward to a complete abandonment of cultivation by workers engaged in the non-farm sector would be disappointed to witness that their interest in agriculture was on the rise. Prima facie, workers primarily occupied in casual non-agricultural labour have been squeezing out agricultural labour and animal farming and focusing their energies on cultivation.

A similar tendency was shown by traders who have shown more interest in cultivation in 2020 than in 2013, abandoning animal farming substantially. Salaried employees, too, have opted for cultivation as a secondary occupation in greater numbers than before. Over three-fifths of all three non-farm category workers have opted for cultivation as a second profession. The message is loud and clear. Despite a massive transformation of the rural occupational structure, where over two-thirds of workers have chosen non-farm work as a priority, cultivation is here to stay, if not as a primary occupation, then as a secondary one.

In contradiction with the assumptions of mainstream transformation theories, virtually all rural workers have maintained a foothold in agriculture. Those prioritising non-farm employment for potentially higher income continue to engage in cultivation. Even if a spike in the reporting of cultivation as a secondary occupation is ignored, cultivation has not shed its ground.

Moreover, a clear occupational break, the much-touted movement of workers from primary to secondary or tertiary sectors, is nowhere to be seen. Workers easily flock between sectors but never forget to keep a foot on the ground.

The rural occupational structure was undergoing massive gender segregation, with agricultural occupations becoming a female domain and non-agricultural employment male bastions. Nowhere was this more evident than in cultivation, with the waning of the proverbial *Kisan*, and female farmers taking over. This phenomenon has often been referred to as the feminisation of agriculture, but it was the "agriculturalisation of female occupation". This point will become clearer once data on other agrarian occupations are considered. But before that, it is worth noting that male participation in cultivation was shrinking, providing space for the expansion of female involvement. However, this process was set in motion earlier; for the first time, the share of female workers occupied in agriculture surpassed that of males by a wide margin. The seven-year period between the surveys witnessed a massive gendered overhauling of the occupational structure in rural India, with a shifting of cultivation towards females.

In another piece of evidence of gendered segregation of occupations, male workers have completely abdicated animal farming. However, females' share in the profession was also subsiding; they were the ones who were running the show. Agricultural labour, too, has become a female occupation, with women covering the ground vacated by male workers. Considering all three agrarian occupations, female participation has already exceeded the 70 per cent mark. They appeared to be filling the space created by the withdrawal of male workers to pursue relatively lucrative jobs elsewhere.

While women continued to be clustered in low-productivity agrarian vocations, male workers moved to high-productivity off-farm jobs. By the time of the second survey, well above two-fifths of them were already in casual non-farm labour. During the same period, the share of female workers in the occupation had come down to single digits. Trading was the only occupation where the gender gap was relatively low. However, like all other occupations, the share of male and female workers in trade was reciprocal, giving credence to the gender-segregation hypothesis. A slightly different pattern was seen in much-coveted salaried employment, where male and female workers increased their presence.

It was not just a coincidence that the share of male and female workers was moving in a reciprocal fashion, except for agricultural labour and salaried jobs. The withdrawal of women from the workforce because of the economic prosperity argument appeared weak in the face of evidence showing gendered segregation facilitating men's movement towards higher productivity jobs and women filling the space left behind.

Not surprisingly, this data also reveals the nature and character of transformation overshadowing the rural. Migrant casual labourers in cities have been considered the urban underbelly, the most exploited in extremely insecure and underpaid jobs. And this capitalist exploitation in urban spaces gets subsidised by gendered labour relations in rural areas. The migrant men labourers couldn't have survived producing for capital in urban areas and reproducing in rural areas without transferring the burden to women.

Alongside gender, caste also rendered services to capital by segregating workers, mediating mobility, and consequently supplying a cheap and controlled labour force. While the Dalits, mostly agricultural labourers and poor peasants in rural areas, essentially became casual labourers, landed UCs opted for salaried jobs. The capitalist penetration in the countryside had destroyed the productive capacities of the rural economy to the extent that migrants were not left with any bargaining power in urban industries.

5 Market, Caste, and Land

The rural transformation essentially encompasses the transformation of land relations. The latter forms the basis of the former. The sociological perspective on land relations believed that the transition towards a modern society entailed a change in land relations. The social system, the economy, and the political structure in rural areas were embedded in land relations. Land relations were often viewed from narrow economistic considerations, sometimes focused on productivity, leaning towards the interests of landlords and rich farmers, and sometimes sacrificing efficiency in favour of the eradication of poverty. However, the sociological perspective did not find a contradiction in efficiency and equity if a long-term perspective informed views on land relations and the entire society (Burman et al., 1974).

Further elaborating on the sociological perspective on land relations, Burman et al. (1974) argued that:

> history tells us that whichever country postponed the abolition of feudalism through land reforms only achieved incomplete, albeit perverted, modernisation…land reform is linked up with the political goal of a democratic society; and whether you are going to have a truly broad-based democratic society, depends on what efforts you make to change the feudal agrarian structure.

Burman et al. further argued that the transformation of land relations was meant to enhance productivity by transferring control from unproductive landlords to productive cultivators.

In a labour-surplus country like India, where men compete for land and not the other way around, Burman et al. (1974) cautioned against the inherent dangers of a relapse of older land relations. Joshi also decried those hopefuls who believed that market dynamics alone would correct anomalies in land relations, citing the overvaluation of land owned for security, status, prestige, rental income, and livelihood. The sociological perspective demolished the artificial compartmentalisation of rural and urban areas by underlining the continuity between the elite and the poor in both locations. While the middle class in the initial years of independence was composed largely

DOI: 10.4324/9781003588245-5

of landlords and individuals from rich farmer backgrounds, rural poverty spilled over into urban areas (Burman et al., 1974).

The sociology of land relations has consistently emphasised its close correspondence with social structure, particularly caste and gender, constituting the axes of domination and dispossession. It could be particularly witnessed in the case of women and Dalits who worked on land much harder than others but remained land-poor quite often. The land was not just another economic asset; in a society deeply rooted in agrarian relations, privileges and deprivations were ingrained into it, even for those who have diversified outside agriculture. The rural social hierarchy and power structure continued to be entangled in feudal culture several decades after the formal abolition of rent-seeking landlords. In a society where the influence of archaic ethos over social relations has not been obliterated, disparity and control formed the basis of relationships – ideologically and materially – and the challenge to inequality has been rather feeble. The paraphernalia of inequality finds sustenance in the close correspondence between land, caste, and gender, and no transformation was possible without uprooting it (Mohanty, 2021).

Breaking this union of "land with relations of domination was one of the objectives of the transformation of land relations. Freeing the land from formal political superfluities allowed it to acquire a pure economic asset. The separation of land as a factor of production from landed property was essential for it to be freely traded in the market. Since the land co-constituted the social structure along with caste and gender, its transformation paved the way for a change in social relations" (Namboodiripad, 1984, p. 7).

Thus, the "commodification of land paved the way for investment by a capitalist farmer by paying rent to the landowner and wages to the hired labourer in order to earn profits. The process went on to replace landlord–tenant-type old relations with modern classes of the landowner, industrial capitalist, and wage-labourer seeking rent, profits, and wages from capitalist agriculture. The production for the market distinguished capitalist agriculture from subsistence farming" (Namboodiripad, 1984, p. 7). However, this long-drawn process acquires various intermittent forms and might not reproduce the experiences of the West in a developing country (Patnaik, 2007; Akram-Lodhi, 2010). These complexities force one to investigate emerging land relations to assess whether the relationship of land with relations of domination and subjugation, particularly caste and gender, was changing. It also demanded a probe into changing patterns of landownership and consequent land concentration or lack thereof.

Another dimension of the transformation of land relations rests on equity and efficiency planks. Agriculture economics in India has witnessed an engaging "inverse farm size productivity" debate that provided an economic rationale for land reforms demonstrating higher productivity by smaller farms compared to larger farms. However, Ghatak and Roy (2007) argued that the persistence of the inverse relationship could be due to frictions in the land market that did not allow larger holdings to break into smaller holdings.

It has also been pointed out that the inverse relationship was relevant only in premodern agriculture, which was completely changed by the application of capital and technology in post-green revolution agriculture (Patnaik, 2007). Ghatak and Roy (2007) also believed that the parties could design tenancy contracts to incentivise the tenant to the extent of an owner-cultivator. If it was not happening, this could again indicate friction in agricultural land tenancy markets.

What these "frictions" could be? Whether the land has not been completely commodified or whether the transformation continues to have imprints of social relations marked by caste and gender. Any study on rural transformation could not escape these questions. This chapter seeks to unravel these issues based on longitudinal data collected from the selected villages. The exercise involves identifying patterns of changes in landownership structure, along with sizes of holdings, caste, and gender. This will include the study of land transactions, both leasing and sale.

The Historical Legacy of Land Relations

Of the four villages under study, land in Senapur and Gohanakallan was completely controlled by Thakurs at the time of independence. Opler and Singh's (1952) survey of the village in the early 1950s revealed that Thakurs directly controlled 82 per cent of land in Senapur as landlords and had an interest in the remaining part of it in some or other forms. In a nutshell, the entire land in the village was under the control of Thakur landlords. All other farming castes were dependent on Thakurs for leasing land. Some of them, like Yadavs and Noniyas, often found their names in revenue records as tenants and were bestowed with some stability and claims. Many others, particularly Jatav-Chamars and other lower castes, were tilling oral tenancies without any right over cultivated land, leaving them completely dependent on landlords. However, after the abolition of Zamindari in the state in the mid-1950s, most of the recorded tenants got ownership rights of land under possession in a long-drawn process stretching over two decades. But oral tenants did not have any claim. Some of them owned tiny plots of land; others waited for more than two decades until the government decided to allot *pattas* (decree of allotment of land parcels) in the mid-1970s. These families till the same not-so-fit-for-cultivation plots located at far-flung places, and Thakurs controlled large parts of the village. At the same time, peasant *jatis* constituted the middle of the agrarian structure.

Another Thakur-dominated village, Gohanakallan, has evolved differently. It resembled Senapur in the 1950s due to the complete hold of landlords on village land, but proximity to a major city, Lucknow, made all the difference. While the impact of land reforms was somewhat similar in both villages, Thakurs maintained dominance alongside the emergence of a middle peasantry composed of Yadavs and Dalits at the bottom of the landownership hierarchy, largely landless or owning tiny plots and *patta* land allotted by the state from the 1970s onwards. The location of the village on the outskirts of Lucknow city galvanised real estate activities in the village from the early 2000s.

While Senapur continues to witness a dormant land market, Gohanakallan was marked by a hyperactive market. A huge chunk of agricultural land has been sold ostensibly for housing purposes. A new locality named Sainik Nagar has emerged on village land, but more than that, the motive behind the land purchase by urban capital was speculation. Economists have always been worried about the dormant land market in rural India that they blame for the anomalies in landownership structure and the maintenance of the status quo in the land-caste equation. Gohanakallan provides an opportunity to determine whether a speculative market could help.

As against Gohanakallan, which the city has completely swallowed, Pandari existed in relative isolation, even in terms of connectivity by road. Leaving aside an active land market, the agrarian economy was undeveloped, given infertile land and the absence of irrigation facilities. However, the incidence of landlessness was lower in the village compared to other villages under study. Brahmins and Thakurs constituted an old class of landowners, while Yadavs and some other backward *jatis* acquired landownership due to zamindari abolition independence. All the villages have one thing in common: the landless and land-poor group was largely composed of Dalits, particularly Chamars. In Pandari, Koris also contributed to this segment.

However, Seemli has an entirely different history of land revenue systems than Senapur and Gohanakallan, but the landownership pattern was marked by the dominance of one caste, in this case, Jat. But one difference that stood out over similarities was that in the western UP village, the land was owned by self-cultivating farmers. It might have positively impacted productivity, but it had adverse consequences for Dalits' landownership possibilities. In Zamindari areas, the abolition of intermediaries, conferment of rights to tenants, imposition of a ceiling, and distribution of government land could transfer a substantial quantum of land from erstwhile owners. At the same time, none of these possibilities existed in the ryotwari (bhaichara) areas. In the latter areas, except for formal change, the situation on the ground did not change noticeably. The entire land reforms architecture was designed keeping zamindars in mind, while self-cultivators were protected.

The limited overhauling of land relations left a legacy of huge landlessness in the entire state, its extent higher in ryotwari areas than in zamindari areas. Irrigated lands, as in the case of Seemli, were also prone to a greater extent of landlessness. The higher productivity incentivised the dominant classes to concentrate land in their hands. Irrespective of legal sanction, the landownership structure was far from equal in villages falling under the zamindari and ryotwari areas. Like the zamindari revenue system, in the ryotwari system, a set of intermediaries, institutionalised or otherwise, existed, and these groups of households formed the upper crust of the peasantry. While it was an institutionalised hierarchy in the former, the social order was rooted in huge land inequality in the latter. Besides various classes of the peasantry, a class of landless labourers was maintained in both areas.

Another commonality that united all the villages as they existed during the survey was an absence of disjunction from the older set of land relations. In all the villages, the relationships that existed in the colonial period were modified after independence. The formal status of groups has changed, but inequality between them has carried forward and exuberated. Identifying a group of landlords among the village households was not difficult. Senapur and Gohanakallan had a group of Thakur families, whereas Jats in Seemli and Brahmins in Pandari formed the core group of landowners.

Further, it doesn't surprise an observer that most landowners, big or small, were legacy owners. The historical continuity of land relations was the most complicating factor in the study of rural transformation. The untangling of land and social relations that have been woven together over a long period could be challenging work for social science research.

Changes in the 2010 Decade: The Landownership Structure

This chapter is based on data collected on agricultural landownership in all four villages by household surveys conducted in 2013 and 2020. The analysis presented in the chapter is based on data obtained from the households common to both surveys. The agricultural land reported in both surveys also included land owned for orchards, etc.

In 2013, nearly 28 per cent of households in Gohanakallan did not own agricultural land (Table 5.1). The share of landless households was 28 per cent in Pandari, 46 per cent in Senapur, and 56 per cent in Seemli. However, in government policies and official statistics, landlessness is considered as owning nil land, including agricultural and homestead plots and land held for other purposes. Because of that, often, the extent of landlessness remained hidden. But when landlessness is defined based on the ownership of agricultural land, the actual extent of landlessness comes to the fore.

Besides very high landlessness in terms of agricultural land, marginal holdings of less than 2.5 acres constituted 65 per cent of all holdings in Gohanakallan, 55 per cent in Pandari, 49 per cent in Senapur, and 32 per cent in Seemli. Smallholdings, defined as owning areas between 2. 5 acres to 5 acres, constituted 6 per cent in Gohanakallan, 12 per cent in Pandari, 3 per cent in Senapur, and 6 per cent in Seemli. The share of ownership holdings covering an area of 5–10 acres, designated as medium holdings, was less than 1 per cent in Gohanakallan, 3 per cent in Pandari, 2 per cent in Senapur, and 6 per cent in Seemli. The large holdings, owning more than 10 acres, constituted 1.7 per cent in Pandari and 0.4 per cent in Seemli. In the remaining two villages, there was no large holding.

Absolute landlessness, defined in terms of nil agricultural land, appeared to have come down in 2020 compared to 2013. In Senapur, it came down by 2.3 percentage points. A corresponding fall was recorded at 8.5 percentage

Table 5.1 Distribution of Households by Land Category (Acres) in Different Villages – 2013 and 2020

Land Category (acres)	*Gohanakallan*		*Pandari*		*Senapur*		*Seemli*		*Total*	
	2013	2020	2013	2020	2013	2020	2013	2020	2013	2020
Absolute landless (<0.1)	117	118	84	72	165	152	134	115	500	457
	27.92	28.37	28.38	24.16	46.09	43.68	56.07	47.52	38.11	35.05
Marginal (0.1–2.5)	274	268	163	165	175	177	77	99	689	709
	65.39	64.42	55.07	55.38	48.88	50.86	32.22	40.91	52.51	54.37
Small	24	27	35	43	11	11	14	17	84	98
(2.5–5.0)	5.73	6.49	11.82	14.43	3.07	3.16	5.86	7.02	6.4	7.52
Medium	4	3	9	11	7	8	13	10	33	32
(5.0–10.0)	0.95	0.72	3.04	3.69	1.96	2.3	5.44	4.13	2.52	2.45
Large	0	0	5	7	0	0	1	1	6	8
(above than 10.0)	0	0	1.69	2.35	0	0	0.42	0.41	0.46	0.61
Total	419	416	296	298	358	348	239	242	1,312	1,304
	100	100	100	100	100	100	100	100	100	100

Source: The Rural Transformation Survey (2013, 2020).

points in Seemli and four percentage points in Pandari. In Gohanakallan, the extent of landlessness slightly increased by 0.45 percentage points.

While the share of marginal landholdings of less than 2.5 acres has seen variations across villages and no prevailing pattern in all four villages, in Senapur and Gohanakallan, their share came down by four and two percentage points, respectively. In the remaining two villages, it went up by one percentage point in Pandari and ten percentage points in Seemli.

One point that came up quite clearly was that historically, the landownership structure of each village was unique, and it was evolving differently. The overall (inclusive of all four villages) share of a sub-category of marginal holdings in the range of 0.6 acres to 1 acre might have shown some stability during the reference period, but wide variations existed across villages. In Senapur and Seemli, it went up by three and 2.5 percentage points, respectively, while in Pandari, it came down by one percentage point, and in Gohanakallan, it remained almost stagnant.

Similarly, in the another sub-category of marginal category (1–2.5 acres), the average of all four villages has slightly gone up, but at the village level, changes varied across villages. In Senapur, the share of holdings of this class recorded an increase of 3.5 percentage points, while Seemli witnessed a fall of equal magnitude. Both the remaining villages retained a share of this class of holdings in 2020, as they did in 2013.

On average, smallholdings have slightly increased by about one percentage point. The share of smallholdings remained the same at both points in Senapur, while Seemli, Pandari, and Gohanakallan witnessed an increase of one percentage point, 2.5 percentage points, and 0.7 percentage points, respectively.

The share of medium holdings was estimated to be similar in 2013 and 2020. On this account, Senapur and Gohanakallan witnessed slight upward and downward movements, respectively, while the share of medium holdings in Seemli and Pandari decreased by 1 per cent and 0.5 percentage points, respectively.

Large holdings experienced a slight increase in their share in 2020 compared to 2013. In Senapur and Gohanakallan, no large holdings were recorded in 2013 or 2020. The share of this class of holdings was stagnant in Seemli at both points, while in Pandari, a slight increase was estimated.

Land and Caste

Table 5.2 gives the social group and land class category-wise distribution of households in the village Senapur of Jaunpur district in eastern Uttar Pradesh in 2013. It presents an alarmingly high level of landlessness among SC households, as around two-thirds of them had no agricultural land. Another 37 per cent of the SC households owned less than 2.5 acres of land, falling in the marginal category. There was only one small and no medium or large holdings among SCs.

Table 5.2 Distribution of Households by Social Group and Land Holding in Senapur – 2013 and 2020

Social Group/Land Category	*Absolute Landless*		*Marginal*		*Small*		*Medium*		*Total*	
	2013	2020	2013	2020	2013	2020	2013	2020	2013	2020
SC	123	112	74	80	1	2	0	0	198	194
Within Social Group (%)	62.12	57.73	37.31	41.24	0.51	1.03	0	0	100	100
Within Land Category (%)	74.55	73.68	42.28	45.19	9.09	18.18	0	0	55.31	55.75
ST	1	1	0	0	0	0	0	0	1	1
Within Social Group (%)	100	100	0	0	0	0	0	0	100	100
Within Land Category (%)	0.61	0.66	0	0	0	0	0	0	0.28	0.29
OBC	39	36	66	62	2	2	0	0	107	100
Within Social Group (%)	36.45	36	61.63	62	1.87	2	0	0	100	100
Within Land Category (%)	23.64	23.68	37.71	35.02	18.18	18.18	0	0	29.89	28.74
Non-OBC–SC–ST	2	3	35	35	8	7	7	8	52	53
Within Social Group (%)	3.85	5.66	67.31	65.99	15.38	13.21	13.46	15.09	100	100
Within Land Category (%)	1.21	1.97	20	19.77	72.73	63.64	100	100	14.53	15.23
Total	165	152	175	177	11	11	7	8	358	348
Within Social Group (%)	46.09	43.68	48.86	50.86	3.07	3.16	1.96	2.3	100	100
Within Land Category (%)	100	100	100	100	100	100	100	100	100	100

Source: The Rural Transformation Survey (2013, 2020).

Against this, 36.5 per cent of OBC households were estimated to be landless, and another 61.6 per cent fell into the marginal category, having less than 2.5 acres of agricultural land. There were only two households of this social group in the smallholdings category and no medium or large holdings.

The landownership pattern of the non-SC–ST–OBC group was entirely different from that of SCs and OBCs. Over 67 per cent of households in this group owned land in the marginal category, but most of them within the range of one to 2.5 acres. Another 15 per cent owned agricultural land between 2.5 acres and 5 acres, while 13.5 per cent of households in this group owned 5–10 acres of land. Absolute landlessness was below 4 per cent in the case of this group.

Further, close to 75 per cent of the landless category was composed of SCs, while the remaining 24 per cent consisted of OBCs, with almost negligible participation from the non-OBC–SC–ST group. The marginal category was also composed largely of SCs at 42 per cent, followed by OBCs at 37 per cent and non-SC–ST–OBC at 20 per cent.

Against this, the medium category was composed of the non-SC–ST–OBC group, while the same group contributed 73 per cent of small landholdings. Only 9 per cent of the total smallholdings were owned by SCs, while OBCs owned 18 per cent of holdings in this category. Marginal-3 had the highest participation from non-OBC–ST–SC at 45 per cent, followed by 35 per cent OBCs and 20 per cent SCs. In a nutshell, the three lower categories were predominantly SC and OBCs, while the three upper categories were preponderant by the "general" category.

The incidence of landlessness among SCs of Senapur came down by around five percentage points in 2020 compared to 2013. Still, around 58 per cent of SC households were estimated to be landless at the latter point. The drop in landlessness has swelled the marginal category by almost a similar magnitude. The 2020 Rural Transformation Survey found two smallholdings among SCs of Senapur as compared to one in 2013. There was no medium or large holding owner household among SCs of the village.

The incidence of landlessness among OBCs of Senapur remained at 36 per cent on both points. The share of OBC households falling in the marginal category has also remained stagnant. About 2 per cent of OBC households were found in the smallholdings on both points.

The data revealed that the SC group experienced some mobility within landless and marginal categories. A fraction of them appeared to have moved from one category to another within the two lowest categories. However, despite a slightly changed landownership pattern, the acute marginalisation in terms of huge landlessness and concentration in lower landholding categories persisted during the reference period.

This is confirmed by the fact that the composition of the landless category remained almost intact on both points. This category was constituted by 74 per cent of SC households in 2020, with less than a 1 per cent fall compared to 2013, while OBCs constituted 24 per cent and the non-OBC–SC–ST composed less than 2 per cent.

Within marginal holdings, the share of SCs went up by three percentage points, while the share of OBCs and non-OBC–SC–ST came down by 2.7 percentage points and 0.3 percentage points, respectively. In smallholdings, while OBCs kept their share intact on both points, non-SC–OBC–ST lost nine percentage points share to SCs. The medium-holding category was completely dominated by non-SC–OBC–ST in 2013 as well as 2020.

Seemli village in Muzaffarnagar district of Western Uttar Pradesh, as shown in Table 5.3, was no exception to the general pattern of high landlessness among SCs. Over 46 per cent of households belonging to the social group did not own any agricultural land. Another 47 per cent of households found a place in the marginal category on account of owning less than 2.5 acres of agricultural land, while only 4 per cent and 2 per cent of households found a place in the small and medium landholding categories, respectively.

Close to two-thirds of OBC households were reported to be landless in 2013. The extent of landlessness was higher among OBCs than SCs because the former were largely composed of Muslims, who have been revealing much higher landlessness. The remaining one-third of OBC households were scattered all over the landownership pattern, making a single-digit presence in all categories.

There was no non-SC–OBC–ST household in the village.

Two-thirds of the landless category was composed of OBC households. It was a different scenario compared to other villages. Backward Muslims were the most marginalised group in terms of landownership. The other one-third of the category was constituted by SCs.

Marginal was divided between SCs and OBCs in the ratio of 58 and 42. In the small and medium categories, OBCs' share was 71 and 85 per cent, respectively.

In 2020, the incidence of landlessness among SCs had fallen to 28 per cent, a steep 18 per cent decrease. However, landlessness among OBCs persisted at the same level of 62 per cent in 2020. The share of SC households in the marginal category increased by 17 percentage points. SC households finding a place in the small and medium categories taken together in 2020 was almost like in 2013.

The share of OBC households categorised in marginal categories on account of ownership of agricultural land increased by two percentage points at the expense of higher categories. In summary, landlessness among OBCs has persisted at the same level, and the share of their households in higher land categories has come down during the reference period.

The composition of the landless category changed in 2020 as compared to 2013 with an increase in the share of OBC households by nine percentage points and a comparable fall in the share of SC. During the same period, the share of SCs in the marginal category increased by eight percentage points, accompanying an equal fall for OBCs. SCs increased their share in

Table 5.3 Distribution of Households by Social Group and Land Holdings in Seemli Village – 2013 and 2020

Social Group/Land Category	*Absolute Landless*		*Marginal*		*Small*		*Medium*		*Large*		*Total*	
	2013	2020	2013	2020	2013	2020	2013	2020	2013	2020	2013	2020
SC	44	28	45	64	4	6	2	1	0	1	95	100
Within Social Group (%)	46.32	28	47.36	64	4.21	6	2.11	1	0	1	100	100
Within Land Category (%)	32.84	24.35	58.44	64.64	28.57	35.29	15.38	10	0	100	39.75	41.32
ST	1	0	0	1	0	0	0	0	0	0	1	1
Within Social Group (%)	100	0	0	100	0	0	0	0	0	0	100	100
Within Land Category (%)	0.75	0	0	1.01	0	0	0	0	0	0	0.42	0.41
OBC	89	87	32	34	10	11	11	9	1	0	143	141
Within Social Group (%)	62.24	61.7	22.38	24.12	6.99	7.8	7.69	6.38	0.7	0	100	100
Within Land Category (%)	66.42	75.65	41.55	34.34	71.43	64.71	84.62	90	100	0	59.83	58.26
Total	134	115	77	99	14	17	13	10	1	1	239	242
Within Social Group (%)	56.07	47.52	32.22	40.91	5.86	7.02	5.44	4.13	0.42	0.41	100	100
Within Land Category (%)	100	100	100	100	100	100	100	100	100	100	100	100

Source: The Rural Transformation Survey (2013, 2020).

the smallholding category by seven percentage points but lost five percentage points in the medium category.

In Pandari village of Chitrakoot district in Bundelkhand, as per Table 5.4, around 39 per cent of SC households were landless in terms of agricultural landownership. The comparable figures for OBC and non-SC–OBC–ST were 22.4 per cent and 13 per cent, respectively.

Close to 53 per cent of SC households, 62 per cent of OBC households, and 41 per cent of non-SC–OBC–ST were categorised as marginal. The highest proportion of the share of a group in the smallholding category came from non-SC–OBC–ST at 23 per cent, followed by 14 per cent of OBC and 6 per cent of SC. A similar pattern was observed in the medium holding category, where 10 per cent of non-SC–OBC–ST households found a place compared to only 1.6 per cent of OBC and 2.3 per cent of SC.

Quite clearly, a larger share of the non-SC–OBC–ST group found a place in the higher landholding category compared to the two other groups.

The data on the composition of the land class categories revealed the marginalisation of SCs and OBCs. The landless category was constituted by about 61 per cent of SC households, followed by 33.4 per cent of OBCs and less than 6 per cent of non-SC–OBC–ST households. In contrast, the medium category was composed of over 44 per cent non-SC–OBC–ST households, 33 per cent SC households, and 22 per cent OBC households. The small category had the largest contribution from OBC households at 51 per cent, followed by non-SC–OBC–ST at 26 per cent and SCs at 23 per cent. The stark inequality in the ownership of agricultural land was most visible in this village.

In 2020, the absolute landlessness among SCs and OBCs had come down by five percentage points each but increased by 2.5 percentage points in the non-OBC–SC–ST group. Against this, the non-OBC–SC–ST and OBC groups had increased their share in the medium category by ten and five percentage points, respectively, at the SC group's expense. In the smallholding category, SCs and the non-OBC–SC–ST group witnessed their share shrink marginally and go to OBCs. There was no large holding in the village in 2013, but seven large holdings were reported in 2020, with six owned by non-OBC–SC–ST groups.

About 42 per cent of SC households of Gohanakallan were landless in 2013 on account of owning no agricultural land. The comparable figures for OBC and non-OBC–SC–ST groups were 17 per cent and 32 per cent, respectively. Another 55 per cent of SC, 75 per cent of OBC, and 53 per cent of the non-OBC–SC–ST group found a place in the marginal category. Less than 2 per cent SC, around 6.5 per cent OBC, and approximately 29 per cent non-OBC–SC–ST households owned 2.5–5 acres (small) land. A fraction of SC and OBC households also owned medium-category holdings.

As with all other villages, the landless category was largely (56 per cent) constituted by SC households, as revealed by Table 5.5, followed by OBC (32 per cent) and the non-OBC–SC–ST group (13 per cent). The composition

Table 5.4 Distribution of Households by Social Group and Land Holdings in Pandari Village – 2013 and 2020

Social Group/Land Category	*Absolute Landless*		*Marginal*		*Small*		*Medium*		*Large*		*Total*	
	2013	2020	2013	2020	2013	2020	2013	2020	2013	2020	2013	2020
SC	51	44	69	77	8	9	3	2	0	0	131	132
Within Social Group (%)	38.93	33.33	52.67	58.34	6.11	6.82	2.29	1.52	0	0	100	100
Within Land Category (%)	60.71	61.11	42.33	46.66	22.86	20.93	33.33	18.18	0	0	44.26	44.3
ST	0	0	0	1	0	0	0	0	0	0	1	1
Within Social Group (%)	0	0	0	100	0	0	0	0	0	0	100	100
Within Land Category (%)	0	0	0	0.60	0	0	0	0	0	0	0.34	0.34
OBC	28	22	77	76	18	24	2	3	0	1	125	126
Within Social Group (%)	22.4	17.46	61.6	60.32	14.4	19.05	1.6	2.38	0	0.79	100	100
Within Land Category (%)	33.33	30.56	47.23	46.06	51.43	55.81	22.22	27.27	0	14.29	42.23	42.28
Non-SC–ST–OBC	5	6	16	11	9	10	4	6	5	6	39	39
Within Social Group (%)	12.82	15.38	41.02	28.21	23.08	25.64	10.26	15.38	12.82	15.38	100	100
Within Land Category (%)	5.95	8.33	9.81	6.66	25.71	23.26	44.44	54.55	100	85.71	13.18	13.09
Total	84	72	163	165	35	43	9	11	5	7	296	298
Within Social Group (%)	28.38	24.16	55.07	55.38	11.82	14.43	3.04	3.69	1.69	2.35	100	100
Within Land Category (%)	100	100	100	100	100	100	100	100	100	100	100	100

Source: The Rural Transformation Survey (2013, 2020).

Table 5.5 Distribution of Households by Social Group and Land Holdings in Gohanakallan Village – 2013 and 2020

Social Group/Land Category	*Absolute Landless*		*Marginal*		*Small*		*Medium*		*Total*	
	2013	2020	2013	2020	2013	2020	2013	2020	2013	2020
SC	65	65	85	87	3	3	2	1	155	156
Within Social Group (%)	41.94	41.67	54.7	55.37	1.94	1.92	1.29	0.64	100	100
Within Land Category (%)	55.56	55.08	31.02	32.46	12.5	11.11	50	33.33	36.99	37.5
ST	0	0	0	1	0	0	0	0	1	1
Within Social Group (%)	0	0	0	100	0	0	0	0	100	100
Within Land Category (%)	0	0	0	0.37	0	0	0	0	0.24	0.24
OBC	37	38	163	160	14	14	2	1	216	213
Within Social Group (%)	17.13	17.84	75.40	75.09	6.48	6.57	0.93	0.47	100	100
Within Land Category (%)	31.62	32.2	59.48	59.70	58.33	51.85	50	33.33	51.55	51.2
Non-SC–OBC–ST	15	15	25	20	7	10	0	1	47	46
Within Social Group (%)	31.91	32.61	53.1	43.48	14.89	21.74	0	2.17	100	100
Within Land Category (%)	12.82	12.71	9.12	7.46	29.17	37.04	0	33.33	11.22	11.06
Total	117	118	274	268	24	27	4	3	419	416
Within Social Group (%)	27.92	28.37	65.45	64.42	5.73	6.49	0.95	0.72	100	100
Within Land Category (%)	100	100	100	100	100	100	100	100	100	100

Source: The Rural Transformation Survey (2013, 2020).

of the marginal category was also along similar lines, with a higher share of SC (31 per cent) and OBC (59 per cent) than non-OBC–SC–ST (9 per cent). The category of smallholdings comprised 58 per cent of OBC households, followed by 29 per cent of non-OBC–SC–ST households and 12.5 per cent of SC households.

The high share of landless households among SCs persisted at the same level in 2020 as in 2013, at 42 per cent. Incidence of landlessness in terms of ownership of agricultural land continued to be at the same level among OBCs (17 per cent) and non-OBC–SC–ST group (32 per cent). The proportion of OBC and SC households in the smallholding category remained at the same level, but the share of the non-OBC–SC–ST group in the smallholding category went up by seven percentage points.

Land Possession and Caste

In the total operated area of Senapur in 2013, SCs possessed 28 per cent, OBCs 31 per cent, and the non-OBC–SC–ST group 41 per cent (Table 5.6). The share of OBCs in Seemli was 63 per cent, and that of SCs 35 per cent. In Pandari, the leading group in land possession was OBC, which cultivated 43 per cent of the total area, followed by 29 per cent of the non-OBC–SC–ST group and 28 per cent of SCs. In Gohanakallan too, OBC had the largest share of the operated area at 62 per cent, followed by SC at 23 per cent and the non-OBC–SC–ST group at 15 per cent.

Overall, the survey data showed that OBC was operating the maximum area in all the villages except Senapur. All four villages taken together, this group possessed 50 per cent area, followed by 28 per cent of SC and 22 per cent of the non-OBC–SC–ST group.

SC households of Senapur increased the area under their possession by eight percentage points in 2020 compared to 2013 (Table 5.7). In Pandari and Gohanakallan, this group gained seven percentage points in each village during the reference period. Only in Seemli village did the SC group lose four percentage points share in the operated area. The OBC group's share in the operated area came down in three villages by 2–5 percentage points. Only in Pandari did their share remain stagnant. The non-OBC–SC–ST group also lost three percentage points share in Senapur, nine percentage points share in Pandari, and four percentage points share in Gohanakallan.

The average of all four villages revealed that the SC group had the highest share of their operated area from leasing at 22.6 per cent, followed by OBC at 14 per cent and the non-OBC–SC–ST group at 8 per cent in 2013. Those groups owning smaller landholdings were more dependent on leased-in land.

In Senapur, around 42 per cent of the total area under the possession of SCs came from leasing. Comparable figures for Pandari were 32 per cent, followed by Gohanakallan (13 per cent) and Seemli (3.5 per cent). At an aggregate level, the share of leased-in land was over 15 per cent of the total

Table 5.6 Distribution of Operated Area by Social Groups and Villages – 2013 and 2020

Social Group/ Village	*Senapur*		*Seemli*		*Pandari*		*Gohanakallan*		*Total*	
	2013	2020	2013	2020	2013	2020	2013	2020	2013	2020
SC	28.39	36.18	34.52	30.79	28.01	35.70	22.96	29.38	28.10	33.17
ST	0.00	0.00	0.00	0.09	0.49	1.49	0.08	0.07	0.21	0.43
OBC	31.01	26.64	62.70	58.93	42.60	42.91	62.42	60.34	50.08	46.22
General	40.59	37.19	2.77	10.19	28.91	19.90	14.55	10.22	21.61	20.18
Total	100.00	100.00	100.00	100.00	100.00	100.00	100.00	100.00	100.00	100.00

Source: The Rural Transformation Survey (2013, 2020).

Table 5.7 Percentage of Leased-in Land in Total Operated Area – 2013 and 2020

Social Group/Village	*Senapur*		*Seemli*		*Pandari*		*Gohanakallan*		*Total*	
	2013	2020	2013	2020	2013	2020	2013	2020	2013	2020
SC	41.83	44.22	3.46	3.47	32.22	41.93	13.11	30.10	22.57	33.32
ST	–	–	–	0.00	66.67	85.00	0.00	0.00	60.38	77.89
OBC	13.94	21.57	1.20	9.70	23.40	22.27	14.64	15.38	14.26	16.77
General	7.49	5.08	0.00	23.89	10.50	7.49	1.06	0.00	7.85	6.66
All	19.26	23.63	1.93	9.22	22.46	27.28	12.33	18.12	15.37	20.48

Source: The Rural Transformation Survey (2013, 2020).

operated area. It was 22 per cent in Pandari, 19 per cent in Senapur, and 12 per cent in Gohanakallan.

In 2020, the share of leased land in the total operated area increased by five percentage points at the aggregate level. This was due to an increase in this proportion across villages. In Senapur, the share of the leased-in area in the total operated area increased by four percentage points, Pandari witnessed a surge of five percentage points, and Gohanakallan recorded an upward movement in the area under tenancy by six percentage points. On average, the SC group reported an increase in the share of tenanted land in the total operated area by 11 percentage points. The increase was two percentage points for OBC. The only non-OBC–SC–ST group recorded a drop in the area under tenancy.

The Land Market and Quality of Land

In the share of the irrigated area to the total cultivated area, the variation could be seen across social groups and villages. Except for Pandari, all other villages had a relatively high share of the area under irrigation from either public or private sources. SCs had a lower share of irrigated land than other social groups. At an aggregate level, 95 per cent of the total operated area was under irrigation.

In 2020, the share of irrigated land in the total operated area reported in all four villages increased by two percentage points compared to 2013. SCs gained four percentage points, followed by non-SC–OBC–ST with three percentage points, and OBC one percentage point. Still, the lowest share of irrigated land in the total operated area belonged to SCs, while OBCs were slightly better off than them. Non-OBC–ST–SC had almost total land under irrigation.

During the seven years between the first survey in 2013 and the second survey in 2020, 77 households in all four villages purchased land. These households constituted about 4 per cent of the total households. The highest number of households purchased land in Gohanakallan (29), Seemli (23), Senapur (16), and Pandari (9).

In all four villages, a total of 72 households purchased agricultural land, while 15 bought homestead land (Table 5.8). In Gohanakallan, 27 households purchased agricultural land, while seven bought homestead land. Comparable figures for Seemli were 20 and seven, respectively. Sixteen households in Senapur purchased agricultural land, and one household bought a homestead plot. In Pandari, nine households bought agricultural land, while no household purchased homestead land during the reference period.

The data on the number and proportion of households that sold land in the seven years between 2013 and 2020 shows that 94 households sold land during the reference period. These households accounted for about 5 per cent of all households in four villages. The highest number of seller households was in Gohanakallan at 49, constituting 9 per cent of surveyed households

Table 5.8 Total Agricultural Land Area (acre) Purchased and Sold by Households between 2013 and 2020

District	*Purchased Land (acre)*	*Sold Land (acre)*
Senapur	14.135	1.307
Seemli	21.478	17.713
Pandari	7.171	6.79
Gohanakallan	22.772	55.997
Total	65.556	81.807

Source: The Rural Transformation Survey, 2020.

in the village, followed by 33 in Seemli (9 per cent), seven in Pandari (2 per cent), and five (0.8 per cent) in Senapur.

In terms of area bought and sold during the reference period, Gohanakallan tops the list with a sold area of 56 acres and a bought an area of 22.78 acres. Seemli recorded a purchased area of 21.4 acres and a sold area of 17.7 acres. The comparable figures for Senapur were 14.1 acres and 1.3 acres. Pandari had the smallest area bought and sold.

All in all, it came out from the data that the land market was most active in Gohanakallan and Seemli, both in terms of the number of transactions and the area transferred. Since the entire area bought and sold was not equal, it was evident that the parties outside the village were also involved in land transactions. For instance, the area sold in Gohanakallan was 33 acres higher than that bought by village households; the land might have been transferred to persons outside the village. However, the area bought in Senapur and Seemli was higher than the area sold; village residents might have purchased land from parties outside the village.

Pandari had almost equal area bought and sold. The land market appeared sluggish in the village as the number of transactions and the area transferred were quite low.

The Homestead land market was sluggish in all four villages. Table 5.8 shows that a very small area of land was transacted during the seven years between 2013 and 2020.

In Senapur and Gohanakallan, marriage expenses emerged as the foremost reason for the sale of land, while in Pandari it was medical expenses (Table 5.9). Seemli households had reported "other" reasons for the sale of land. The construction of houses and the purchase of another piece of land were other important reasons.

Concluding Remarks

The data presented in the chapter shows some variations during the reference period. Notwithstanding some variations across land size classes, the major picture was kept intact. Dalits continued to constitute two-thirds to three-fourths of landless households in villages, except Seemli, where OBC

Table 5.9 Reasons for Sale of Land in Different Villages (Percentage)

Reason/Districts	*Senapur*	*Seemli*	*Pandari*	*Gohanakallan*	*All*
Marriage Expense	25	3.03	14.29	31.11	19.1
Education Expense	0	6.06	0	0	2.25
Medical Expense	25	12.12	42.86	8.89	13.48
Debt Payment	0	12.12	28.57	4.44	8.99
Other Business	0	6.06	0	2.22	3.37
Purchase Land	0	6.06	0	22.22	13.48
Legal Battle	0	0	0	2.22	1.12
Construct House	25	3.03	0	24.44	14.61
Other	25	51.52	14.29	4.44	23.6
Total	100	100	100	100	100
(*N*)	(4)	(33)	(7)	(45)	(89)

Source: The Rural Transformation Survey, 2020.

Muslims formed the larger part of the category. They were overwhelmingly represented in the marginal category, essentially functional landless owning less than half an acre of land.

On the other hand, non-SC–ST–OBC continued to be preponderant in higher land categories in three villages. This landowning class was primarily constituted of Thakurs in Gohanakallan and Senapur, and Brahmin-Thakur combined in Pandari. Interestingly, the fourth village, Seemli, appeared to have two poles within the OBC; between two-thirds to three-fourths of the contribution in 2013 and 2020 to the landless category came from the social group, and 85–90 per cent of the medium category holdings also belonged to the intermediate group. The OBC group in Seemli contained two major segments – Muslims and Jats; the former was preponderant in the bottom category, while the latter clustered in the larger category.

Another relatively stable trend was the share of a social group in land size classes in an ascending or descending order. The proportion of Dalit households was distributed descendingly, highest in the bottom category and gradually decreasing as one ascended land size classes. It was the other way around in the case of the non-OBC–SC–ST group. OBC was found to be the most internally diverse category and hence did not show any pattern of landownership spectrum. One would often find the share of OBC households distributed across land size classes.

The structural status quo in the land-caste equation, albeit with some flux, draws from the state's failure to dismantle old relations. In place of breaking the impasse through the vigorous implementation of land reforms, the modified version of colonial land relations was allowed to thrive. For instance, all landowning Thakur households in Senapur are descendants of village zamindars. The same is the case with Gohanakallan Thakurs, whose holding size might have come down due to generational subdivision; collectively, they owned all lands that their forefathers controlled as sir and khudkasht at the time of zamindari abolition. The two other villages were also not an exception to the general pattern of land contained within the caste of old landlords.

The market also did not break this impasse. Pandari and Senapur witnessed negligible households participating in the land sale market. However, the purchased area in Senapur was not as insignificant as Pandari, but considering the size of the village, it only denoted a sluggish market. In Seemli and Gohanakallan, the number of participating households and quantum of the transferred area were estimated to be substantial during the seven-year reference period. The similarities between both the villages do not go beyond this. Households of Gohanakallan reported having purchased about 23 acres of land and sold around 57 acres, indicating that people from outside the village bought a larger share of land for housing purposes. Qualitative data revealed that most of them were city dwellers who invested in the village, considering its proximity to the town. End-users did not make these purchases, but a variety of intermediaries and small-time developers did. The urban capital's interest in the landed property was solely for speculative purposes. In due course, the urban would swallow the rural.

In the not-so-distant future, the urban capital's encroachment on the landed property of the village will be complete, leaving behind what is often termed the "urban village". The dynamics of speculative market transactions in the land have been enriching the landed classes of the village, impoverishing the landless and the land poor, who were fast losing whatever access they had to the land.

The quantum of land purchased by Seemli households was higher than the area they sold. These were mostly intra-village transactions. However, Seemli happened to be connected to the urban area, but unlike Gohanakallan, land deals in the village were for agricultural purposes. They were essentially inter-landholding transactions, but the bought and sold area was insignificant to dent the overall landownership pattern. However, these land transfers delivered the land to cultivating households from non-cultivating ones or those willing to quit agriculture.

The land sale and tenancy market data revealed a close relationship. Since both these markets transferred the land between cultivating households and non-cultivating households, activities in one could dampen activities in the other. For instance, around one-fourth of the operated area in Senapur and Pandari in 2020 was leased-in land. These were the two villages where the land sale market was sluggish. At the same time, tenanted land constituted less than one-tenth of the operated area in Seemli, where significant land sale transactions took place during the reference period. Gohanakallan appeared as an exception where many land sale deals coincided with a considerable quantum of the area under tenancy. But a closer look at the qualitative data would reveal that Gohanakallan was passing through an alteration. One part of the village was already under the influence of real estate agents, while agricultural activities persisted in the other. The weight of pre-existing land relations and the urban capital was reflected in the interesting data supporting our hypothesis of one market neutralising the other. One must admit that, in some cases, the two markets overlapped when agricultural land purchased by the real estate agents was leased-out for farming. Usually, this happens during the period when investors wait for the escalation of the landed property price.

However, some participants in land sale or tenancy markets carried more strength than others. And the power to determine the market was rooted in the landownership structure. Some caste groups were more dependent on leased-in land than others. For instance, more than two-fifths of the area operated by Dalits in Senapur was tenanted land, while UCs leased barely 7.5 per cent of cultivated land. Similarly, leased land constituted one-third of the operated area of Dalits in Pandari, whereas it was only one-tenth for the UCs. The caste-land equation described above continues to be strong despite heavy leasing.

The data allowed one to see that Dalits in Senapur constituted 55 per cent of the households but owned less than 17 per cent of the area in 2013. Whereas UCs of the same village, with 15 per cent share in the number of households, cultivated over 37 per cent of the area. Pandari was no different. Dalits contributed over 41 per cent of households and 19 per cent to owned land. The Dalits of Seemli were estimated to be relatively better placed with around 40 per cent share in households and 31 per cent in owned land. With a proportion of 37 per cent in the number of households, the Dalits of Gohanakallan had a 20 per cent share in owned land.

However, the story didn't end in 2013. It continued to show divergence between the landownership structure and the pattern of the operated area, the latter hyperactive to compensate for the rigidity of the former. Though the landownership structure persisted in a similar shape till 2020, Dalits increased their share in the possessed area in all the villages except Seemli, mainly at the cost of non-OBC–SC–ST. They augmented the share by eight percentage points in Senapur, seven percentage points in Pandari, and six percentage points in Gohanakallan. During the same period, the non-SC–ST–OBC group and OBCs in Senapur lost three and five percentage points of area, respectively. The loss was exclusive in Pandari and primarily in Gohanakallan to non-SC–OBC–ST. OBCs maintained area under their possession in both the villages. During the intervening period, the substantial area transfer was via the tenancy market in the face of a sluggish sale market. The variation in the share of leased land in the operated area of Dalits and the non-OBC–ST–SC group testified to it. It appeared that UCs were holding ownership of the agricultural land but leasing it out to Dalits.

Whether the increasing distance between the class of landowners and the class of cultivators could be termed the return of semi-feudalism? The sluggish land market and the proliferation of small tenancy indicated the absence of transformation of land relations. Primordial relationships like caste continued to have a grip over land relations. Drawing such a conclusion would be presumptuous. It would be based on the premise that small tenancy and rent-seeking landowners existed only in a semi-feudal political economy. This judgement flows from an inability to distinguish between the land tenancy of older land relations and its modern incarnation.

The sluggish market does not necessarily imply that land has not been commodified. Social and political relations have a close interface with landownership in rural areas, but these equations increasingly materialised via the market and democratic set-up. For instance, Gohanakallan has seen the emergence of a class of neo-rich from among the old landlords through the real estate market. Old landlord families did not hesitate to alienate the land retained within the caste for generations once real estate activities shored up prices. Several land sale transactions in rural areas are conducted through caste networks, often within the caste, but the land was sold at market-determined prices.

The contemporary tenancy is not a replica of the old-style small tenancy. However, the powerful party dominated the tenancy market, be it the lessee in reverse tenancy (as in Seemli) or the lessor in small tenancy, but new forms of agricultural land tenancy were not obligatory but contractual. What we understand from "small tenancy" was not necessarily landless or marginal peasants leasing land from a large owner but often "parallel" lease contracts between similarly placed landowners. Often, decisions regarding leases were made by both parties considering the availability of labour in the household in the context of a closer integration of the agrarian economy with the market economy.

The land relations in the countryside are transforming at a pace, but they could not break from the past.

6 Agriculture, Productivity, and Technology

In the debate on rural transformation, agriculture has always occupied a special place to the extent that the transformation of agriculture was considered synonymous with the transformation of the rural, effectively equating "agriculture" with "rural". But a rupture in the bonhomie between "rural" and "agriculture" is increasingly becoming difficult to ignore. This point has been discussed theoretically in the Introduction in Chapter 1 and empirically in the occupation and employment chapter that rural transformation, while closely linked with agrarian transformation, is not synonymous with the latter. The contention is that agrarian transformation necessarily triggers rural transformation, but the former is not a prerequisite for the latter in the contemporary political economy.

Before that, one intends to explore the trajectory of conceptual overlapping between rural transformation and agrarian change. The most glaring example of the linkages between the two could be discerned from Breman and Mundle's (1991) Rural Transformation in Asia, consisting of contributions on agrarian change. It wasn't incidental that a book on rural transformation largely focused on agrarian change. The first chapter by Byres (1991) clears doubts about the work's intent by describing the role of agriculture in the transformation of the overall economy by generating an investible surplus for industrialisation in an economy's initial stage of development. It wasn't possible without achieving a degree of productivity on the basis of the transformation of agrarian relations. Needless to say, under these conditions, a drive from the agrarian change could transform the rural.

But contrariwise, this might not be necessarily true. The rural transformation could be achieved without compulsorily transforming the entire agricultural sector. The closer integration of the rural economy with the capitalist market economy offers this opportunity. In this situation, agriculture could be relegated to playing a pillion to the industries and service sector where most of the rural workers would be occupied. The "non-agricultural sector" would likely mean extremely insecure jobs on construction sites, workshop manufacturing, and petty trading. We have argued elsewhere that these informal conditions of employment do not allow the transformation of agriculture to proceed beyond a certain point. Workers continue to maintain a toehold in the farm sector as a shock absorber, something their employers

DOI: 10.4324/9781003588245-6

in the town never offered. The absence of the social security umbrella by the state only meant that a clear break from agriculture was not an option. These two processes have often been referred to as "push" and "pull", and transformation from the below or from the above.

Both of these processes might appear conceptually distinct, but empirically they could often be intertwined. Incompletely transformed agriculture might continue to support other sectors of the economy by providing food grains to the burgeoning urban population (Myrdal, 1968) at below-cost prices, in turn facilitating industries to hire "cheap" labour. It is no hidden secret that the family labour of farming households mostly goes unwaged and unpaid. This pauperised agrarian economy would be unable to unleash rural transformation, which the market economy would eventually trigger by partially absorbing rural workers. One calls it partial absorption because migrant labourers form an expendable workforce located at the industry's periphery. They are employed for a limited period of time without any job security and written contracts; they earn low wages and lack any sense of social security. These migrant workers survive at a bare minimum in towns and frequently plough back incomes to villages, boosting demand for industrial goods and services and scaling up the rural market. This two-way movement is facilitated by increasing connectivity between the rural and urban areas where the state plays no small role.

The relative contribution of both a push from the below and a pull from the above might yield qualitatively different results. When workers are "released" from developed agriculture, armed with state-provided education, health, housing, and food security, they would have stronger bargaining power in the labour market. On the contrary, when the capitalist market encroaches upon the rural economy, the net result would be destitute workers without a say in their employment conditions. Between these pole positions, a household's location in the agrarian structure essentially determines prospects in the urban labour market. Needless to say, the agrarian sector in Uttar Pradesh, much like elsewhere, is far from homogenous, and the transition of households from one sector to another is largely horizontal in the near absence of state support for social development.

The penetration of capital in the rural economy also destroys crucial linkages between different subsectors, further distancing rural transformation from agrarian change. Capital deals directly with subsectors, transforming them differently. Otherwise, two major employment-generating subsector – the farm and rural non-farm sectors – could grow in mutually interdependent ways. In that condition, the character of these linkages would depend upon the nature of growth in the agricultural sector. Medium and small-farmer-led growth was found to be more conducive for the non-farm sector, rather than rich-farmer-centred growth. For the same reason, the implementation of redistributive land reforms has been positively correlated to the development of the rural non-farm sector, which was no longer considered a residual, transitory phase. Sections of workers have often been found rotating between the sectors across seasons (Bhalla and Chadha, 1983; Chakrabarti and Kundu, 2009).

Under both conditions, the shifting of labour from agriculture to non-agricultural sectors has rightly been identified as the major aspect of transformation. It was also reflected in changing occupational patterns of the working-age population discussed in Chapter 4. In order to assess the contribution made by the agrarian economy to the transformation of the rural, two aspects have been discussed in this study – land relations that form the basis of agrarian relations in Chapter 5 and the present chapter dealing with changes in the agricultural economy. As discussed above, the underlying factors of the level of food grain production were crucial to the transformation of the macroeconomy as well as the rural economy.

Agricultural Changes

This study looks at the transformation from both sides. But this chapter is particularly concerned with the changes experienced by the agriculture sector, which employs the vast majority of rural workers. This chapter's most crucial question is about the change of Indian agriculture from a low-productivity subsistence farming plagued by pre-capitalist forms of production relations into high-productivity, capitalist agriculture. The movement from a subsistence activity to a market-oriented enterprise involves the adoption of desired cropping patterns, achieving considerable yield levels, and, in turn, higher production, culminating in a marketable surplus. These objectives could not be achieved without the adoption of newer technologies in seeds, farm implements, etc. The chapter analyses data on all these aspects of agriculture in studied villages. The chapter's primary objective was to assess the direction of change; hence, aggregate data have been presented. However, disaggregated analysis has been referred to in the text whenever required.

Before one gets down to the processes at the micro level, taking a macro view of the economy would set the backdrop. Agriculture in Uttar Pradesh is undergoing a transformation. The state has witnessed a massive increase in food grain production. The surge in volume has been attributed to productivity in the context of almost stagnant (Singh et al., 2021), or shrinking, agricultural area. The total agricultural land in the state came down from 17,983,000 hectares in 2006 to 17,450,000 hectares in 2016. The fall of about 533,000 hectares shown by agricultural census data was believed to be diverted to non-agricultural purposes, including infrastructure and housing.

The state's growth in food grain production is synonymous with the rising production of wheat and rice. Both crops have been expanding their area at the cost of other crops and are experiencing productivity gains. However, recent studies have underlined the plateauing yield of wheat. In terms of cropping pattern, the state is heading towards Punjab-style wheat-rice monocropping. The foremost casualties of this shift are pulses and coarse cereals, which are losing area despite increased productivity, resulting in declining production (Singh et al., 2021).

The transformation of agriculture might be assessed using data on cropping patterns, production and yield levels, and adoption of technology, all three closely interlinked with each other. Studies in agricultural economics have often focused on determinants of productivity enhancement. For instance, irrigation and usage of fertilisers have emerged as the consensus candidates for drivers of productivity across regions and crops (Pandey and Reddy, 2012). However, as Singh et al. (2021) pointed out, many of these studies were premised on a linear growth pattern model and simply focused on technical inputs at the cost of social factors.

Besides fertiliser and irrigation, it is not surprising that cropping intensity enhances productivity by way of increasing gross area when the potential of the net area expansion has been almost saturated (Singh et al., 2021). Otherwise, the conceptual relationship between the cropping intensity and productivity ensures that gains made in the first are necessarily reflected in the second. In intensive agriculture, the problem of adherence to strict timelines and a tiny window between two crops often leaves farmers reliant on farm machines, implements, and controlled sources of irrigation. By way of precise input application and staying ahead of the crop cycle, agriculture's mechanisation has been positively correlated to productivity (Singh et al., 2021).

Singh et al. (2021) have also brought social attributes into the spotlight of productivity analysis. They show that rural literacy was a critical element in preparing farmers for the adoption of new agricultural practices. Studies have also shown that traditionally self-cultivating peasant caste households have proved to be more efficient (Trivedi, 2011). On the contrary, rent-seeking upper caste (UC) households were much of a drag on the development of agriculture. The massive mechanisation spree after the 1960s changed everything. Armed with machines, a number of resident UC households have not only withdrawn leased-out lands but have been looking for land to lease (Srivastava, 2016). Alongside, the emergence of the farm implement lease market has enhanced the reach of small and marginal farmers to hitherto inaccessible capital assets; however, not without certain constraints. Both these changes together have set in motion a process of convergence in the yield levels of crops, except for some commercial crops where rich farmers still enjoy a considerable advantage. But the same cannot be claimed in the case of men-owned farms and women-owned farms, the latter being generally impoverished (Agarwal, 1994).

Cropping Pattern

The cropping pattern provided the key linkage between agrarian change and rural transformation. Both the processes intersected at the site of the non-farm sector since the two sectors communicated with each other through forward and backward linkages. The supply-side and demand side linkages reflected the growth in agriculture, translating into a boost in demand for the non-farm sector. In turn, the non-farm sector provides a market for agricultural produce and supplies technical inputs to the latter (Chakrabarti and Kundu, 2009).

The production of basic crops, particularly food grains, has consistently supported a strong relationship between the two sectors in rural areas. Chakrabarti and Kundu (2009, p. 71) argued that a:

> small farm economy specialising in non-mechanised production of basic crops generated on one hand substantial agricultural employment and hence large market for cheap mass consumption goods, and on the other, demand for locally produced and supplied inputs of production and farm services.

The policy push often perceives cropping patterns as a single-handedly subjective decision, but human intervention is just one of the many factors responsible for the cropping pattern of a particular location that evolved in the context of agro-climatic conditions, quality of soil, sources of irrigation, and market availability. Climatic conditions and soil types supportive of a variety of plants have been historically the most crucial. While rains and natural water bodies have been there all along, human-made sources of irrigation made the first change. Harnessing groundwater and channelling surface water made it possible to grow hitherto impossible crops. However, the entry of the market in this triad turned everything upside down. Cropping patterns that evolved over time got unsettled under the impact of market penetration in a relatively shorter span. Market demand often leads to dramatic distortions of existing patterns, frequently leading to an economic, social, and ecological crisis. However, the entire agriculture might not be market-oriented, but a market-dominated ecosystem influences subsistence farming too (Kundu and Chattopadhyay, 2018).

Often the notion of "crop diversification" has been deployed in the probes on changing cropping patterns. It essentially indicated the cultivation of various crops on the same plot, maximising production and minimising biological and market-related risks. But the underlying assumption was the movement from low-value crops to high-value ones with the assistance of new technology (Kundu and Chattopadhyay, 2018). However, the data from two rural transformation surveys witness the notion of crop diversification pulling apart in two different directions. The movement towards high-value crops might bring uniformity rather than diversification. The issue of the price sensitivity of Indian farmers has been a settled question for a long time.

The dynamics of crop diversification have been held responsible for risking micro food security. While macro food security could still be achieved via the institution of the market, the movement of farming families away from food grain cultivation threatens food security at the household level. Besides, this shift leaves the non-farm sector vulnerable, which has been dependent upon the agricultural sector for food grain supply. Cultivation of high-value crops might require modern implements and inputs, sourced from urban manufacturers, displacing the rural non-farm sector. Farm machinery, inputs, and other industrial consumer goods have also flooded rural markets as part of the process where agriculture seeks to fulfil the demands of urban consumption (Harriss, 1991; Sen and Raju, 2006; Chakrabarti and Kundu, 2009).

Besides capitalist markets, state intervention is no less crucial in catalysing changes in cropping patterns. The supply of subsidised inputs, procurement at support prices, science and technology innovations, and infrastructure for storage and irrigation, among other things, propel changes in a policy-desired direction (Kundu and Chattopadhyay, 2018). The weightage accorded by the state's policies to different crops might again reduce the variety of crops. The massive support for wheat and rice since the 1960s has pushed out a range of millets and other crops from farmers' fields and people's food (Kumar, 2019). Punjab is often cited as an example in this debate (Mann, 2017).

The cropping patterns of the studied villages were no exception. The impact of state intervention and market integration could be easily discerned from the longitudinal data. The data on the cropping pattern of surveyed villages in 2013 (Table 6.1) revealed that the largest sown area during the Kharif season was under Paddy in Senapur (158 acres) and Gohanakallan (332 acres). Pandari prioritised Kharif pulses by devoting 101 acres, followed by coarse grains (76 acres) during the same season. Seemli was almost entirely covered by the annual crop of Sugarcane (180 acres).

But during the Rabi season, cropping patterns across villages looked almost uniform, with wheat occupying the largest area. Pandari and Gohanakallan devoted 250 acres each for wheat in this season, while Senapur's area under wheat was 144 acres in 2013. Besides wheat, Pandari also cultivated Kharif pulses (41 acres), oil seed (42 acres), and coarse grain (36 acres). Oil seeds were also sown in Senapur (13.2 acres) and Gohanakallan (37 acres). Senapur also cultivated coarse grains on 8.8 acres and sugarcane on six acres. Coarse grains were also grown in Seemli on seven acres of land.

The Second Rural Transformation Survey, 2020 also collected data from households on cropping patterns. The diversity of cropping patterns appeared to have shrunk during the seven years. For example, in the Kharif season, the 2020 survey recorded only four crops, namely, paddy, pulses, corn, and vegetables, compared to six crops recorded in 2013, namely, paddy, pulses, coarse grains, groundnut, and vegetables. In addition to the annual crop of sugarcane, the Rabi season had five crops in 2013, viz wheat, pulses, oilseed, coarse grain, and vegetables. It came down to four crops seven years later: wheat, pulses, oilseed, and coarse grains.

The relative preference between crops has not changed substantially during the period. Paddy continued to command the highest area during the Kharif season in Senapur (263 acres) and Gohanakallan (263 acres). However, Pandari only had pulses (46.6 acres), dropping coarse grains, oilseeds, and

Table 6.1 Cropping Pattern (Land Sown in Acre) in Four Villages: 2013 and 2020

Crops/Districts	*Senapur*		*Seemli*		*Pandari*		*Gohanakallan*		*Total*	
	2013	2020	2013	2020	2013	2020	2013	2020	2013	2020
Paddy	157.62	263.22	21.83	0.00	8.78	6.00	331.85	263.04	520.08	532.25
Kharif Pulses	11.99	21.41	1.75	0.80	100.85	46.59	1	0.00	115.59	68.79
Corn	–	4.31		0.00		0.00		0.67		4.98
Coarse Grain	9.89	–	2.26	–	75.796	–	0	–	87.946	–
Groundnut	2.52	–	0	–	16.2	–	0	–	18.72	–
Oilseed	0.1	–	0	–	30.17	–	0	–	30.27	–
Wheat	144.22	265.95	51.81	70.48	250.54	363.64	250.57	282.81	697.13	982.88
Rabi Pulses	0.74	3.95	0	0.00	41.25	40.15	0.75	2.10	42.74	46.20
Rabi Oilseed	13.21	9.24	1.28	0.50	43.45	67.80	37.44	1.80	95.38	79.34
Coarse Grain	8.76	14.12	6.93	0.58	36.22	1.20	0.3	0.00	52.21	15.91
Sugarcane	6.06	4.57	179.94	196.66	0.3	0.00	0.72	0.00	187.02	201.23
Vegetables	5.37	81.32	0.96	0.92	0.25	0.00	20.46	12.57	29.08	94.80

Source: Rural Transformation Survey (2013, 2020).

groundnut from its cropping pattern. The area under vegetables has increased substantially in Gohanakallan. Senapur did not cultivate coarse grains and groundnuts in 2020, instead shifting slightly to corn. Seemli continued to be under sugarcane during both seasons.

The "big-four" – paddy, wheat, sugarcane, and vegetables – seemed to have emerged victorious at the aggregate level. Each of these has expanded its area previously occupied by other crops during the seven-year period. Wheat made the longest stride, while coarse grains, groundnut, Kharif pulses, and oil seeds have been experiencing displacement. Clearly, the cropping pattern was the net sum of forces applied by the state policies, market demand, and other institutional-agro-climatic factors. It would not escape anybody's attention that wheat, rice, and sugarcane are the three crops that occupy the highest priority in government intervention in agriculture. Most importantly, wheat and rice receive support in the form of a policy of minimum support price (MSP). Sugarcane prices are covered by a fair and remunerative price (FRP) regime. However, groundnut, pulses, and oilseeds are also covered by the MSP regime, and some of these crops receive additional attention from the government; still, they appear weak in front of the mighty wheat–rice–sugarcane triad.

The government claims that the MSP of all the crops is fixed at 50 per cent over and above the cost of cultivation (CoC). But in the case of wheat, this additional component has been much higher than in other crops lately. A completely overwhelming presence of wheat could be attributed to a rather skewed crop parity. Mustard, too, fetches an MSP double the cost of cultivation. According to the government data, the prices of most Kharif crops are also fixed with a 50–60 per cent margin over the C2[1] CoC. Still, paddy maintained an overshadowing, if not controlling, presence during the Kharif season.

The qualitative data from the studied villages showed that ensuring food security at the household level was always on the minds of cultivators. Farmers, big or small, simply didn't like buying from the market, particularly staple diets like wheat and rice. Even when diversifying, they grew these crops on smaller land plots. Secondly, MSP has been a cropping pattern driver, but not beyond a point. In Uttar Pradesh, less than one-tenth of the farmers, mostly resourceful, have access to the MSP system, with the remaining dependent on the vagaries of the market. The farmers try to hedge their bets on a particular crop by ensuring that if the MSP system fails them, they have a backup in the form of a market. It was no hidden secret that the two agricultural commodities traded in the largest volumes have the most widespread market channels down to the level of village traders.

Sugarcane and vegetables have their own dynamics, often grown by cultivators at the top and bottom ends of the agrarian structure, respectively. Sugarcane is a demand-driven commodity that promises high returns based on larger investment and, more than that longer maturing period, stretching over a year. Once sown, it can be harvested more than once with minor irrigation and crop maintenance investment. On the flip side, it was an industry-facing commodity, so the payment cycle might take longer. These institutional constraints ensured that only resourceful farmers who could sustain themselves without income from sugarcane cultivation and manage working capital for a long period endeavoured to farm this crop.

If sugarcane cultivation was possible only in a sugar mill's catchment area, vegetable farming was sustained in the vicinity of a town or a marketplace. These were shorter duration crops ranging from one to three months, required smaller investments, but more tending hands. A typical vegetable farm would be on a tiny plot, often leased in and operated by communities that have been traditionally associated with vegetable vending. However, rural areas near metropolitan cities, for instance, the National Capital Region of Delhi, might draw farmers from a well-to-do background, particularly in potato cultivation. It was one of the commodities whose shelf life could be prolonged by cold storage. The fast perishable green vegetables were still the forte of the first variety of cultivators.

Agricultural Output

The agricultural economics in India has given scant regard to institutional and social factors; however, not without notable exceptions. Thanks to A K Sen's contribution in the sixties, an enduring debate on the relationship between farm size and productivity has overshadowed much of the discipline in more than half a century since then. One strand in this leaned towards an inverse relationship between farm size and productivity, relying on an extensive application of family labour whose imputed value was cheaper to the farmer, in Sen's (1964) opinion, as compared to hired labour to a commercial farmer (Deolalikar, 1981). Economists have argued that land was a perfectly divisible factor (Bardhan, 1984; Ghatak and Roy, 2007).

The other camp in the debate consistently pointed out that the advantage to small farms due to the "self-exploitation" of family labour only existed in "traditional" agriculture and withered with the advancement of technology penetration (Deolalikar, 1981). The data in "green-revolution" areas revealed a positive relationship between farm size and productivity, the latter deploying superior technology for commercial agriculture (Patnaik, 1987). However, one must point out that these studies were conducted soon after the introduction of high-yield variety seeds (HYV), fertilisers, and pesticides in areas of assured irrigation. This period was marked by a limited proliferation of green revolution technologies both spatially and vertically.

Recent studies have presented a mixed bag of evidence (Mishra and Singh, 2021). For instance, Chand et al. (2011) reiterated an inverse relationship for India, contending that smallholders were not lagging in technology adoption, whereas Singh et al. (2021) underlined a significant fall in productivity with an increasing share of marginal and smallholdings in the data. Similar variations have been found across states and crops (Dev, 2012).

The farm size productivity debate has consistently highlighted two factors – family labour and technology; their absence or presence, or a mix of both, produces results for farms (Sen, 1964). Family labour has its own strengths – a free flow of labour, caring, tending, and nurturing plants, and judicious usage of resources. Technology came with a deeper understanding of plant science, precision in the timely deployment of available resources, and above all, a scale of operation almost unthinkable by human endeavour alone. The story around these two poles initially tilted towards small farms that had a reserve pool of family labour, followed by a leap in productivity among large farms that were early adopters of technology, paving the way for a diversified agrarian scene where small farms are equipped with both the factors in a varying degree and large farms have consolidated their hold over the latest technology.

But surprisingly, despite correctly identifying technology and labour as two causes moulding farm productivity, the debate persisted in the framework of the size of the farms. It is quite understandable that the debate, influenced by the land reforms discourse, initially relied on farm size categories. The inverse relationship proposition provided a rationale for breaking down inefficient large holdings, seeking social equality and productivity in food grains. The debate served a good cause but eventually got stuck in the farm-size framework. It could never break from its initial antecedents despite clearly reading the script underlying the phenomenon.

The subject of the proliferation of technology, both spatially and vertically, has been treated in some detail in the next section. Suffice it to add that the assumption of rich farmer-new technology and small farmer-elementary technology has little empirical support. However, this dynamic is not without pitfalls, and an equation skewed in favour of rich farmers. But the bottom line is that small farmers were not completely excluded from the newer technology. Armed with the resource of family labour, the small farms compete with large farms on yield per acre in food grain cultivation. Before one enters into the debate on the role of technology in productivity enhancement, a quick look at the data on total output and yield would be in order.

Coming back to the empirical data collected in the 2013 survey, the total paddy output was 2,167 quintals in Senapur and 3,288 quintals in Gohanakallan. The other two villages had a minor quantity of output. Kharif pulses were grown in Pandari (325 quintals) and Senapur (70 quintals). These two villages also produced coarse grains of 250 quintals and 82 quintals, respectively. The Bundelkhand village and Avadh village also produced groundnut and oilseeds.

During the Rabi season, the total wheat output of Gohanakallan was 2,836 quintals, followed by 2,210 quintals in Pandari, 1,625 quintals in Senapur, and 678 quintals in Seemli. The highest quantity of coarse grain took place in Seemli (1664 quintals), followed by Pandari (130 quintals) and Senapur (45 quintals). The output of oilseed was 176 quintals in Gohanakallan, 130 quintals in Pandari, and 62 quintals in Senapur. Pandari also produced 161 quintals of Rabi Pulses.

One might be surprised to observe the low quantities of produce in Seemli. The answer lies in the quantity of sugarcane produced by the village. The total output of sugarcane in Seemli was 48,859 quintals.

The 2020 survey captured that the output of paddy in the agricultural year 2019–2020 was 4,798 quintals in Senapur and 4,335 quintals in Gohanakallan. The output of wheat in Senapur was 4,320 quintals, followed by 3,628 quintals in Gohanakallan, 3,828 quintals in Pandari, and 1,157 quintals in Seemli.

By the time of the second survey, the sugarcane output in Seemli reached the level of 62,562 quintals in the agricultural year 2019–2020.

Taking all four villages together, the paddy production reached 9,207 quintals in 2020 as against 5,806 quintals in 2013, with an average annual growth rate of 8.36 per cent. During the same period, the aggregate wheat output jumped from 7,348.4 quintals to 12,933.1 quintals. The average growth rate of wheat production turned out to be 10.85 per cent per annum. But the growth rate of sugarcane output was not as impressive. The aggregate sugarcane production rose from 51,427 quintals in 2013 to 63,134.3 quintals seven years later, registering an annual growth rate of 2.64 per cent. Sugarcane production in the western UP village was probably touching its peak at the time of the first survey in the present technological environment. But the overall growth in the agricultural output has been remarkable. Alongside the cropping pattern, it reflects a faster pace of changes experienced by agriculture.

Crop Yield

More than cropping pattern and total output, the yield level was considered a crucial indicator in assessing the transformation of agriculture. Table 6.2 gives the yield levels of various crops in 2013 and 2020.

The average productivity of paddy, including all four villages, in 2013 was 11.2 quintals per acre. It was 13.75 quintals/acre in Senapur, 13.74 quintals/acre in Seemli, 5.77 quintals/acre in Pandari, and 9.91 quintals/acre in Gohanakallan.

The average productivity of wheat in 2013 was 10.54 quintals/acre. The comparable figure was 11.26 quintals/acre in Senapur, 13.09 quintals/acre in Seemli, 8.82 quintals/acre in Pandari, and 11.32 quintals/acre in Gohanakallan. Sugarcane productivity in Seemli was 274.98 quintals/acre.

The average productivity of paddy in 2020 increased by 6.14 quintals/acre compared to 2013. It was a jump of 55 per cent in seven years, indicating a 7.85 per cent growth rate per annum. This increase was highest in Gohanakallan at 6.57 quintals/acre, followed by 4.5 quintals/acre in Senapur.

The average wheat yield in the seven-year reference period increased by 2.6 quintals/acre. It was an increase of 25 per cent between 2013 and 2020

Table 6.2 Yield (Quintal Per Acre) of Different Crops in Four Villages: 2013 and 2020

Crops/Districts	*Senapur*		*Seemli*		*Pandari*		*Gohanakallan*		*Total*	
	2013	2020	2013	2020	2013	2020	2013	2020	2013	2020
Paddy	13.75	18.23	13.74	–	5.77	12.50	9.91	16.48	11.16	17.30
Kharif Pulses	5.82	0.96	3.89	3.02	3.22	2.15	3.00	–	3.50	1.79
Corn		5.93		–		–		4.48	–	5.73
Coarse Grain	8.26	–	6.42	–	3.30	–	–	–	3.94	–
Groundnut	4.58	–	–	–	1.24	–	–	–	1.69	–
Oilseed	6.00	–	–	–	1.39	–	–	–	1.41	–
Vegetables	57.69	–	–	–	2.00	–	51.47	–	25.59	–
Wheat	11.26	16.24	13.09	16.42	8.82	10.53	11.32	12.83	10.54	13.16
Rabi Pulses	5.34	2.92	–	–	3.91	2.43	3.33	29.62	3.93	3.71
Rabi Oilseed	4.67	3.48	5.08	8.00	2.99	0.84	4.70	3.39	3.92	1.25
Coarse Grain	5.14	0.41	240.10	0.98	3.59	0.83	3.33	–	35.24	0.47
Sugarcane	227.23	125.20	277.09	318.13	53.33	–	243.06	–	274.98	313.75
Vegetables	54.00	–	107.29	–	64.00	–	68.04	–	66.61	–

Source: Rural Transformation Survey (2013, 2020).

at the rate of 3.57 per cent annually. The highest rise in yield levels was observed in Senapur, at five quintals/acre, followed by 3.5 quintals/acre in Seemli, 2.07 quintals/acre in Pandari, and 1.5 quintals/acre in Gohanakallan.

The sugarcane yield rose to 313.75 quintals/acre in 2020, reflecting a growth rate of 2 per cent per annum.

The steep rise in the productivity levels of major crops indicated that agriculture in these villages was transforming faster.

Adoption of New Technology

The dynamics of technological adoption in agriculture have been markedly different from industry. Goodman et al.'s (1987) work underlined the "natural production process" of agriculture that acts as a structural constraint against the development of agriculture along the lines of other branches of industry. The authors of this interesting work pointed out that the factory system could not offer a replacement for the conversion of sunlight into food, and this process remained largely biological, signifying the uniqueness of agriculture. They located the distinctive technological transformation of agriculture in this uniqueness of farming.

Elaborating on the distinctive features of agriculture that precluded the possibility of a factory-like system, Goodman et al. (1987) emphasised "biological conversion of energy, biological time in plant growth and animal gestation, and space in land-based rural activities" (p. 1). The imminent failure of technological advancements in clearing natural hurdles to create a "unified production process" paved the way for adaptation to the specifics of agriculture. The new technological strategy relied on "taking over discrete elements of the production process". For instance, machines have replaced animals for ploughing, and organic manure has been replaced by chemical fertilisers. Goodman et al. (1987) further argued that various steps and tasks of agricultural production became a site of technological intervention. The transformation of agriculture was thus the "discontinuous, but the persistent undermining of discrete elements of agricultural process" in place of an overall takeover by the technological transformation (p. 2).

One might not fully endorse the theoretical proposition advanced by Goodman et al., but the approach of looking for elements of agricultural operations that are experiencing technological advancement, while others are only modified in the process, is worth considering. The co-constitution of the agricultural production process by the coexistence of different types of technologies questions the black-and-white narratives of "backward" and "advanced" technologies. It calls for a more nuanced approach to the issue of technological development rather than a static view.

This could be seen in the works of Chakrabarti and Kundu (2009), where the relationship between farm and non-farm sectors has been premised upon an understanding that small farms necessarily relied upon elementary technology. They argued that small farms producing food grains used ploughs, spades, chaff-cutters, sickles, axes, choppers, etc. from the non-farm sector. In a mutually reinforcing equation, the non-farm sector depended on farms for food grains. There is no doubt that the relationship between the two sectors has historically evolved in the ways described by Chakrabarti and Kundu. The farming households sourced simple agricultural implements from blacksmiths, carpenters, etc. and paid them in grains at the time of harvest. But the assumption of small farms and elementary technology in the contemporary rural scene doesn't hold well in the face of mounting evidence to the contrary. The farm size and technology relationship has been at the centre of many debates in agricultural economics and elsewhere in India. These debates have frequently taken the shape of the scale neutrality debate, institutional reforms and technological change debate, and structural change and technology debate.

Dantawala (1970) was one of those who believed in the scale neutrality of the new technology. Dantawala argued that the new technology was not dependent on farm size; instead, it was likely to "soften" dualism – the agrarian structure composed of dynamic capitalist and backward peasant sectors. But technology played out in an institutional framework. Raj (1969, 1970) pointed out that the growth rate of large and smallholdings would differ markedly under the impact of technological progress since the skewed distribution of credit might lead to the concentration of gains from new technology in resource-rich large holdings. But Mellor (1968, p. 358) reasoned that "new technologies were likely to benefit farmers in proportion to the size of landholdings rather than the quantum of labour input" (Joshi, 1974).

The new technology was seen as aggravating vertical differentiation and spatial disparities. As in the green revolution, these technologies were more suitable for regions endowed with fertile soil and abundant water resources, and they hardly had anything to offer to other areas. Mellor (1968) argued that under three conditions, (1) the presence of a large class of landless agricultural labour or tenants, (2) the scarce availability of non-farm jobs, and (3) "income increasing effect was greater than the labour-demand-raising effect of the technology" (p. 358), the income gap between different classes of cultivators was bound to rise exponentially.

A consensus emerged among social scientists that new technology would widen inequalities in the short run; however, some opined that it was likely to ameliorate extreme poverty in the medium term (Mellor, 1968; Raj, 1969, 1970; Joshi, 1974). Hunter (1969) saw ruinous consequences for the peasantry under the new technology regime, to the extent of outright eviction from the land or being forced to sell it off under debt obligations.

Finally, both positions zeroed in on the institutional reforms. Those who believed in the efficacy of technology in delivering growth with justice considered institutional reforms a prerequisite. And others found institutional reforms sans technological progress a non-starter. While land reforms and credit restructuring have often been mentioned among institutional reforms, the transformation of resource and power balance was equally crucial for the realisation of technological gains for the poor, if any (Joshi, 1974).

The great technology debate covered various aspects of the phenomenon, but the nature of new technology received scant attention, both in social sciences and policy. The new technology was presumably capital-intensive. Rao (1965) dissented from framing the debate in terms of labour vs technology. He argued that "labour-intensive" techniques alongside a relatively egalitarian distribution of land would enhance productivity by bringing ownership, operation, and labour on the farm together.

Technology has often been found intertwined with the social structure, specifically gender and caste in India. But many works have taken a highly constricting view of the technology, for instance, confining it to machines and chemicals in agriculture, completely obfuscating the social aspects underlying this process. This narrow view cognitively blacks out the processes that essentially constitute "ways of making things", including the raw materials, commodities, and the socially constituted labour process. A probe into the technology cannot remain confined to mechanical equipment, chemical processes, and biological innovation but must necessarily involve social relations, including gender and caste, that closely interact with the productive processes (Sen, 1985).

The broader view of technology and social arrangements illuminates several hitherto dark spots. For instance, child rearing and family care have been kept in the "cultural" domain despite contributing directly to the economy by ensuring a steady supply of labour. Similarly, the sociology of technology might offer an explanation for the sustenance of power relations between groups around the axis of caste and gender. The distinction between the cultural and economical, the former associated with unpaid and unwaged labour, and the latter paid labour, helps maintain systematic biases that, in turn, keep women economically marginalised (Sen, 1985).

By now, it was no secret that new technology was displacing the women's labour force, thanks to social science research since the 1980s. While these researchers clearly underlined the unfavourable impact of new technology on women workers, they didn't blame the technology itself but held underlying social relations responsible for the differential consequences for different sexes. For instance, Agarwal (1981), on the basis of a study of agricultural modernisation and third-world women, argued that technological innovations per se were not exclusionary but acted in this manner following the socio-political design (Sen, 1985).

Sen, while endorsing this view of social relations being determinant of gendered implications rather than characteristics of the new technology, questions the dynamics of technological innovations. Why would technology veer in the direction of replacing the work that generally women in third-world countries did? For Sen, economic rationality would demand the displacement of higher-paid male workers rather than female workers. Still, the new technology hurt the interests of the lower-paid rural women. The key to this mystery lay in the nature of women's work and the cost of acquiring new technology. The agricultural tasks often assigned to female workers were repetitive and simple, and it did not take much to mechanise them with low-end technology. It made perfect economic sense for the farm owners to replace women's jobs in agriculture with machines at a lower cost (Sen, 1985).

Sen's analysis of putting the onus on the "traditional role" of women, which left the technology unfriendly, giving clemency to the dynamics of technological innovations, could be an interesting question for a deeper probe. One must also ask whether only the mechanisation of women's tasks in agriculture made sense from an economic rationality point of view, and if not, why the consequences differed for different sexes. The present study has explored the question of technological development in detail on the basis of quantitative data from the household survey. And the study collected qualitative data to explore the social consequences of new technology.

During the Second Rural Transformation Survey, the respondents were asked about the adoption of newer technology in agriculture during the seven years preceding the survey. The highest share of households that responded affirmatively was from Senapur at 29.2 per cent, followed by 17.5 per cent in Gohanakallan, 9.5 per cent in Pandari, and 9.1 per cent in Seemli.

However, the lowest share of households in Seemli did not mean that this village was behind all other villages in technology adoption. It could also mean that the Western UP village had already adopted some of the newer advancements in agricultural technology. But at the same time, one could not assume the same thing about Pandari, another village with a lower share of households adopting newer technology. It could be a sign of persisting backwardness in the Bundelkhand village.

The rotavator was the most frequent newer technology adopted by the farming households of the studied villages. Sixty households in Senapur adopted this technology in the seven years between the two surveys. Rotavator Harvester and Cultivator Rotavator were two other agricultural implements adopted by ten and 94 households in the same village, respectively. Seemli witnessed the adoption of cultivators by 15 households, followed by cultivator-rotavator by nine households. Sixty-seven farming households in Gohanakallan focused on the adoption of quality seeds in seven years, while 24 households in Pandari adopted hybrid seeds. Overall, the highest number of households adopted the cultivator-rotavator, followed by quality seeds, hybrid seeds, and cultivators. Three households in Senapur also adopted a harvester combine.

The respondents in the household survey of 2020 were asked about their opinion on the impact of technology adoption on income. On average, over 86 per cent of respondents reported that their income increased after the adoption of new technology, while 11 per cent felt no change. In Gohanakallan, all the respondents experienced the positive impact of technology adoption on income, while the proportion of such respondents was

91 and 92 per cent in Pandari and Seemli, respectively. Surprisingly, Senapur was the village that witnessed a higher number of households adopting new technology, but the lowest share of respondents admitted the positive impact of technology adoption on their income.

When the adoption of new technology was quite a visible phenomenon, diversifying into new crops appeared not to have kept pace with other changes in agriculture. On average, about 2.6 per cent of farming households had adopted a new crop in the seven years preceding the second survey. Senapur was the only village where more than 6 per cent of households reported adopting a crop. In all other villages, the share of such households was negligible. The household data also reiterates that except for Senapur, the area under the new crop was insignificant in all other villages. In Senapur, the area under new crops was almost equally divided between paddy and wheat. It might surprise many, but the adoption of new crops in these villages meant beginning the cultivation of wheat and paddy. The crop "diversification" in these villages was moving along the lines of changes in the cropping pattern of the state that strengthened wheat–paddy monocropping at the expense of coarse grains, groundnuts, Kharif pulses, and oil seeds.

Two of the most important points that emerged from the data on reasons for adopting paddy, wheat, and other crops as new crops in the seven years related to the price-responsive nature of Indian agriculture and the state intervention in the form of price influencers as a determinant of the cropping pattern.

It was quite clear from the data that profit motive was the major reason for the adoption of newer crops in the studied villages. These villages were not an exception but represented a general tendency where Indian farmers were becoming more and more price responsive to draw more profits.

Another noteworthy characteristic of the emerging agrarian scene was the adoption of paddy, wheat, corn, and mustard, all the commodity prices of which were governed by the minimum support price regime of the state. However, the MSP regime only covered government buying to stock subsidised food programmes. Theoretically, market prices were free, but it was no secret that MSP also acted as a benchmark for market transactions.

The data once again reiterated that more than anything else, Indian farmers drew influences from their peer group. More than 86 per cent of farmers who adopted paddy, 84 per cent who adopted wheat, and 85 per cent of farmers who adopted other crops admitted inspiration from other farmers. The data confirms the "peer-group theory" about Indian farmers that might be seen at work in the proliferation of "green revolution" practices across the state.

Finally, regarding the impact on income, the data confirmed what was already reported in earlier paragraphs. It was obvious that almost all the farmers who adopted new crops with profitability in mind had successfully augmented their income on account of new crops.

Concluding Remarks

In a nutshell, the cropping pattern of all the villages was changing. It might not be a true reflection of the notion of crop "diversification" in the sense of cultivating a larger number of crops while moving to high-value produce rather than confining to food grains. A movement within the food grains was visible against the hopes of policymakers and social scientists alike. These villages were no exception to the larger parts of north India, where farmers left behind millet and opted for wheat and rice.

It might be counterintuitive to some people, particularly among urban middle classes, that a big push towards millets, both from the market and the state, has failed to halt a Punjab-like march towards wheat and rice monocropping. While the health-conscious middle class has grabbed market offerings with both hands. "Superfood" is the buzzword of the day. Organically produced food items – millets, vegetables, and fruits – sell at a premium. Governments have announced a promotion policy for millet production.

Actually, the discourse on millets has often been romanticised, calling India the "land of millets" and millet being common people's food while wheat was consumed by the affluent (Kumar, 2019). There is no doubt that before the green revolution, millet cultivation was abundant, and it was an essential ingredient of food. But this argument essentially sets aside the cultural aspirations of the people. They did not eat wheat but relished the idea of it over coarser millets. Those who want to understand the cultural perception of wheat must read the great Hindi writer Prem Chand's short story *Dhaai Ser Gehun*[2] written about a century ago. The widespread opinion of wheat as a smoother grain and good food might have played a role in prioritising it during the green revolution.

Nobody wanted to dispute the benefits of millet cultivation to the health of people and ecology. But the problem was that the green revolution has altered the food choices to the extent that going back to the previous era would be a herculean task. Farmers cultivated to ensure food for the family and to earn cash from the market. It entailed growing crops that they ate and that sold in the market easily. Among other factors, the crop choices were also contingent upon these considerations.

Another factor that weighed on the minds of farmers was the availability of market channels. Wheat and paddy might be classified as traditional food grain crops, but they sold like cash crops. Marketing channels for these commodities in most of north India have reached the doorstep of farming households, however, not without drawbacks. On the other hand, the market for millets has been underdeveloped and demand is nascent. These commodities might be fetching higher prices in urban markets, but trade channels remain exclusively confined to some dynamic farmers.

Interestingly, the yield levels in these commodities did not vary tremendously. And whatever variations in the disaggregated survey data were found, they were more complex than simply small vs large holdings. Perhaps it was the finer balance between labour and technology that yielded the desired results. The qualitative data also revealed that labour relations in general, particularly family labour and access to technology, changed dramatically in the last few decades.

The labour relations in agriculture were marked by higher peak wages and a shortening of the engagement period, thanks to the technological transformation of cultivation. Large swaths of north India were already double-cropped areas, and many regions were progressing towards multi-cropping. Farmers were left with a small window for each operation, and tasks were expected to be completed with utmost precision. This time compression was key to changes in labour demand and wages. It decreased the overall demand for labour but shortened the period of engagement, pushing peak-time wages upwards.

The typical family labour was no longer found. Most working-age male members and a significant number of women would have one foot in the agricultural field and another in a town or elsewhere. These were not the yearlong agricultural workers who returned to the village during major agricultural operations. However, a major feature of the agrarian labour force persisted in the transformed milieu, the difference in the form of engagement between upper-caste cultivators and agriculturalists from peasant castes. While the former continued to perform the role of a farm manager and rarely soiled their hands in the mud, irrespective of landholding size, the latter, while operating the farm, also performed manual labour in the fields. An often-overlooked fact was that the category of family labour was also a caste phenomenon that included backward classes and Dalits. Only among these castes did the practice of exchange labour historically evolve and persist, possibly to a reduced extent in contemporary times.

Another development that might have surprised technology theorists, whose work was based on ownership of new innovations, was the strengthening of the rental market for agricultural machines. The debate on the proliferation of technology was based on the assumption that ownership provided access to innovations, linking it with the size of the holdings. On the flip side, it excluded those who could not afford to buy it. What turned out on the ground was the emergence of a huge rental market for machines and services that brought them closer to hitherto excluded classes of farmers. While the western UP villages commonly followed the ownership model suited for the relatively considerable size of the holdings and level of productivity, the central and eastern farmers with smaller holdings relied on renting machines and irrigation systems. The rental market was relatively underdeveloped in Bundelkhand, as were the instances of ownership.

Beyond this general scenario, the size of the capital asset and the size of the holdings influenced the decision to opt for ownership or rental models. For instance, harvester-combines mostly came on rent, but tractors, threshers, tubewell pumps, sprinkler systems, etc. were among the commonly owned implements. It was true that only the ownership of a machine could ensure the timely performance of a particular task, which culminated in a differential yield. Another important factor that emerged in recent times was the extent to which a particular household was invested in agriculture. Two comparable households might resort to two different strategies depending on their reliance on the agricultural economy and the engagement of family members. Households that were heavily invested in agriculture would frequently opt for ownership despite smaller holdings to encash the opportunities in rental markets.

The premise of large holding-new technology and small holding-simple technology assumed the availability of both regimes. But with the changing rural landscape, for instance, vanishing grazing lands and ponds, the old animal-powered implements were no longer feasible. Migration-centred rural work culture also did not support the rearing of oxen. In short, the ecosystem combining human–animal–land has ceased to exist. Now, it was only the new technology that ruled the agrarian scene.

Rental markets might have bridged the gap in access to technology between classes of farmers, but complete faith in technology and blame at the doors of social relations, as in Sen (1985) and Agarwal (1981), would ignore the politics of technological innovation. Sen's position did not explain why innovations moved in a particular direction. For instance, small farmers would often wonder, looking at the predominance of big machines in a country supposedly belonging to tiny plots. Suffice it to note that the dynamics of innovation were not outside social relations.

Social relations, such as gender, class, and caste, often interacted with innovation and closely corresponded with societal response to new technology, ensuring that the impact was calibrated as per the social location. Sen was mistaken in believing that only women were displaced by new technology. In fact, as Agarwal's work has shown, and the qualitative data of this study also confirmed, the gender-wise differentiated impact was not confined to displacement but also to new opportunity structures. The exclusion from the new jobs was severely felt by women workers.

Finally, it would be an overstatement to term the rural society in India "post-agrarian" (Harriss and Jeyarajan, 2016) where "classes of labour" (Bernstein, 2004; Lerche, 2013) live in place of peasants. The rural might have broken away from agriculture, but not the other way around. Agriculture continued to be firmly rooted in the rural. Rural's break from agriculture has been largely in sources of income and occupation. But it would be far-fetched to conclude that agriculture ceased to be economically relevant in rural workers' livelihoods.

An appreciation for emerging rural-agriculture dynamics would require abandoning a bird's-eye view in favour of a "street view" and asking relevant questions from a worker's perspective. The vantage point of the exploration might be why rural workers maintained their association with cultivation despite a larger share of income coming from non-agrarian sources. The most likely response would be that cultivation provided food grains for the yearly needs of the family. Undoubtedly, a large chunk of cultivators in Uttar Pradesh were subsistence farmers. Another possible response would be that other sources of income have exceeded in share, but agriculture continued to provide income, particularly based on the labour of those household members who did not migrate, seasonally supported by migrant members.

More importantly, the question of livelihood security came up quite forcefully during the second survey in 2020, which began a few months after the national Covid lockdown. It was the time when millions of workers, after losing their jobs, returned home, battered by harsh conditions in towns and penniless long-distance walks. They had reached villages, realising that the job they took to heart did not stay with them even for a day. They were so easily dispensable for their employers, who did not bother to clear their dues. And the state did not make any arrangements for their survival in towns and their journey back home.

Completely abandoned by the state and the market, if something espoused them and gave sustenance, it was a small plot of land in the village. The agricultural sector provided the migrant workers with a sense of security in the face of extremely insecure jobs in the towns and an uncaring state. Tiny plots that contributed hardly one-third to less than half of peasant household income gave them assurance of survival, something they yearned for in relatively better-paid urban jobs. Workers might not have expected a sudden, unprepared lockdown or wholesale job losses, but they always kept a toe in the field to feel secure. The moral of the story was that until urban jobs remain insecure, agriculture would remain pivotal in the livelihood strategy of rural households.

Notes

1 Includes imputed value of family labour, all paid-out costs on inputs like seeds, fertilisers, pesticides, manure, hired human, animal and machine labour, irrigation, depreciation on farm implements and buildings, power charges, interest on working capital and rent on leased-in land. https://www.pib.gov.in/PressReleasePage.aspx?PRID=1868760 downloaded on 16 March 2023.

2 In short, this is a story about a Kurmi peasant named Shankar who wanted to offer relishable food to a visiting saint instead of the barley flour that he had at his place. Shankar loans one and a quarter kilograms of wheat (literally *sava ser gehun*) from a Brahmin of the village who later turned into a moneylender. Shankar tried to repay the loan by working hard, and eventually became a debt-bonded labourer to the moneylender and died after working many years for him. In the end, the Brahmin took Shankar's son as a bonded labourer to repay his father's debt. In the story, Prem Chand referred to wheat as the divine food.

7 Social Inequalities, Market Dynamics, and the State

Although the longitudinal data for the study was gathered from north Indian villages, the problem of stunted rural transformation runs as a common strand across developing countries. Many countries in Asia, Africa, and Latin America face developmental challenges primarily due to the absence of social development paraphernalia in rural areas and almost non-existent formal jobs in urban areas, despite a debilitated agrarian economy. These countries might not be destined to achieve Western-style industrialisation; they cannot afford to escape problems of backwardness, poverty, low levels of labour productivity, and the persistence of social inequality and primordial hierarchies in rural areas.

Historically, the state has played a crucial role in rural transformation in all countries, particularly in the Global North; what sets apart the Global South was the expectation from the state to decolonise the policies and institutions. But the dominanace of neoliberalism in these countries undermines these aspirations by consistently favouring market solutions over state interventions, effectively shaping the rural change. The bigger travesty was that the policies in developing countries continue to be shaped by Western models, seeking sectoral transition despite evidence to the contrary. Against this backdrop, the vital role of social science was to understand the challenges before transformation and, more than that, deconstruct the notion of transformation itself.

The dictum that Europe's past was developing countries' future needed to be shunned first in favour of a nuanced analysis of processes at work. In many developing countries, agriculture remained a large employer despite its declining share in output. The question of developing agriculture could not be dropped from the state's agenda just because the agrarian question has lost centrality in the industrialised world. Countries of the Global South continue to be home to millions of small farms that refused to wither away despite the articulation of the sector within capitalism. Instead of "easing out" the small peasantries, public policy needed to be reoriented towards their progress. These countries could not afford to shift huge number of workers from primary to other sectors in the foreseeable future.

Without an expected expansion of manufacturing jobs, the rural non-farm sector (RNF) has emerged as a crucial site for state intervention. Some studies

DOI: 10.4324/9781003588245-7

have gone to the extent of arguing that RNF would play the same role in developing countries as industries did in developed nations. The sector would absorb a large number of workers released from agriculture. However, these hopes were not completely unfounded, but RNF was not homogenous across countries, and its claim of higher productivity could not be taken at face value. A closely related question of rural–urban migration needed urgent attention from the state in almost all developing countries.

However, these issues have been in the public domain for a long time, but a renewed debate was necessitated on account of the increasing realisation that economies in these countries were not treading the path charted by the industrialised world. For instance, it was quite well known that the rural economy in India was witnessing a separation between agriculture and the countryside. But the challenge was that the share of people dependent on it did not reduce proportionately, adversely affecting their productivity, reaching one-third of urban workers. Conversely, the newly set-up capital-intensive rural manufacturing grabbed a share of the economy but failed due to employment generation (Chand et al., 2017). There was nothing "rural" about these manufacturing units except that they were installed outside the notified urban areas for want of larger space.

These technologically modern manufacturing units hardly offered respite to unskilled women workers displaced by the momentous pace of mechanisation in agriculture, often mistakenly referred to as the "withdrawal" of female workers. The exclusionary nature of the new technology was evident in the disproportionate impact it had on poor women. Construction was the only sector that offered respite, particularly on rural sites in India (Chand et al., 2017).

The peculiarity of the construction boom that arrived with the new millennium was its largely rural character. Initially, people thought construction was happening in urban areas to cater to the demand for expanding corporate offices and mushrooming residential neighbourhoods, and the employment spurt in the countryside only reflected rural–urban migration. It took some time and analytical efforts to realise that instead of urban skyscrapers, millions of tiny rural residential constructions reflected a huge shift in employment patterns. As often happens in developing countries, three major state programmes kickstarted the boom in India, but it went much beyond them (Mahajan and Nagaraj, 2017).

Despite the divergent paths that developing countries seem to have undertaken, they resemble many parts of the now-industrialised nations in dissolving the erstwhile sharp distinction between the rural and the urban. The contemporary rural could better be described as an inferior urban replica. These developments once again question the artificial watertight compartmentalisation between the two locales and partially lend support to the notion of the "rural–urban gradation" and partially to the "collapse of rural–urban categorisation".

While both these propositions dismiss rural and urban as conceptually distinct and separate entities and underline degrees of urbanity and rurality in defining characteristics of a particular locale, the gradation hypothesis, resembling the now out-of-fashion rural–urban continuum paradigm, focuses on the quantitative variations in time and space and the dynamic relationship between types of settlements. In contrast, the new rurality hypothesis probes changes in this equation, presumably under the impact of modernity (Rajagopalan, 1961; Chatterjee et al., 2015, Jodhka, 2016).

The gradation thesis underscores the blurred boundaries between rural and urban areas, rural to urban migration, the expansion of urban amenities in the countryside, and movement of workers out of agriculture, essentially concluding that rural–urban gradation better captures contemporary empirical reality. The subsumption of adjoining villages into cities, the seamless interaction between the two, and the heightened frequency of non-farm activities in the countryside have been cited to buttress the proposition. The proponents of this view successfully tested their hypothesis on gendered variation in the work participation rate and found that the female work participation rate in India decreases along the rural–urban gradation, while no such pattern was underscored for males moving from smaller villages to large cities. The gradation proposition was also found relevant in explaining the availability of regular-wage jobs along the sizes of locations (Chatterjee et al., 2015).

The second strand of the debate stresses the levelling effect of capital on erstwhile rural–urban distinctions appearing in various forms, including the decreasing importance of the primary sector in economic life and the diversification of employment outside agriculture to the extent that it gave rise to the phenomenon of "pluri-active" households. Alongside, the appearance of village markets has changed considerably with the rapid expansion of trade and service establishments in villages dealing in industrial goods and technological services (Jodhka, 2016).

It's a misnomer that urban has reached rural. The converse is also equally true. If the former has knocked on the door of the latter, riding on the wave of the ever-expanding markets, the countryside has overwhelmed the metropolitan with people in various forms and shapes – workers, students, consumers, patients, citizens, and so on. In the process, the rural has expanded its ambit. On one end, it kept on holding agriculture; on the other, it didn't want to be left behind in urbanity. Agriculture might not be the first occupational choice for the younger generation of farmers, but it provided a stepping stone. Moreover, the RNF fills the gap between high hopes and lacklustre deliveries by the urban. A complicated situation like this demands state interventions on the sites of agriculture, RNF, migrant labour, and social development of the rural population.

While exploring these crucial aspects of rural life, this study investigated transformation as a complicated process rather than a predetermined, unilinear process modelled on Western historical experiences. It implied a probe into tension among emerging relations, institutions, and ideas. Diverging from the standard practice of disciplinary compartmentalisation, this study considered economics, politics, sociology, and culture as different facets of rural society and explored their interconnectedness.

This method was crucial in validating earlier understanding that the penetration of market forces would replace pre-existing relations, in Indian cases often premised on caste. This method also allowed for the exploration of the relevance of "modification" and/or "appropriation" in place of complete replacement as a defining process in rural India.

The framework on which the official data edifice relied was also subjected to probing in studying the main questions. For instance, the data sets might be reporting a majority of the people engaged in agriculture, but the fact was that the vast majority of "farming households" had their primary source of income in the non-agricultural sector. This process could not be termed "quitting agriculture" since many workers shifted from agricultural to non-agricultural sources of income while maintaining a toehold in agriculture.

In one more sense, this study departs from other works on transformation. It locates the intersection of *jati* and gender with land as the analytical axis around which the story of transformation has been woven. To be precise, the probe considered economic relations embedded in social ones and vice-versa. The pursual of the longitudinal data in the preceding chapters has made it abundantly clear that caste and gender impinge upon economic changes. Not just social structure, but pre-existing relations, in general, were often modified, appropriated, restructured, and repurposed instead of being outright replaced.

The study begins by posing questions on the characteristics of the transformation underway in India's rural hinterland. The foremost question that comes up in the quest to understand the dynamics of change is who drives the transformation, whether it is industry or agriculture. And whether agriculture is a drag on the momentum of change, in case the transformation is led by industry. In order to assess the nature and character of the transformation, another crucial question would be whether workers employed in agriculture opted for the industry willingly or whether the carrying capacity of the agriculture sector got exhausted.

Further, the questions about the relative importance of linkages between industry and agriculture also come up in the enquiry. Historically, capital, labour, and market linkages have been examined in different times and spaces. But in contemporary India, which of the following would be most crucial: (1) the siphoning off of agricultural surpluses to industry, or (2) the release of workers from the primary sector to other sectors, or (3) the role of the agricultural sector in providing a home market for industrial goods?

While categorising households based on the primary source of income, reporting of agriculture was relatively less frequent. The highest number of households reported non-agricultural labour as their primary source of income.

Demographic Transition and Occupational Mobility

The two closely interwoven aspects of the dynamics overwhelming the rural – the demographic transition and occupational mobility – could be a site for enquiry into the interplay of various factors involved in the transformation. These two aspects acquire significance in the wake of labour linkage between the rural and urban becoming crucial for the contemporary transformation. While both demographic transition and occupational mobility push forward the transformation, restricted mobility was a significant challenge. The missing link of social development that would translate demographic "dividend" into vertical mobility partly explained the phenomenon of restricted mobility.

Demography

Beginning with demographic changes, the four selected villages were located in different regions and exhibited divergent demographic profiles. Senapur, the eastern UP village, appeared to be a Dalit-preponderant village, whereas OBC emerged as the largest group in Seemli. Both SC and OBC social groups had an almost equal presence in Pandari village of Bundelkhand, along with noticeable non-SC–ST–OBC households. Gohanakallan's social group-wise composition slightly differed from the other three villages, with an OBC majority, but the share of SCs was not way behind and included a considerable group of non–SC–ST–OBC households.

Studied villages have been undergoing a fast demographic transition, reflected in the transformation of the age structure in the seven-year reference period. An investigation into the share of different age cohorts revealed that the child population in all four villages decreased gradually. Taking the people of all four villages together, the up-to-14-years cohort had an average share of one-third of the population in 2013, which came down to about one-fourth in 2020. The same data also revealed a slight increase in the share of the elderly in the total population of all four villages. The fastest-growing age cohort in all four villages was 30–59 years, while, as has been clear from the above data, the under-14 group in all four villages recorded a negative growth rate.

The varying growth rate of different segments of the populace also resulted in a changed social and religious group and gender-wise composition. The average share of the SC and non-OBC–SC–ST population slightly increased in all four villages, while the percentage of the OBC population came down. The share of Muslims in the population of Senapur and Seemli, the two multireligious villages, came down slightly alongside a corresponding increase in the Hindu population. The average sex ratio of all four villages dropped significantly in seven years.

The average share of the illiterate population to the total population in all four villages has slightly decreased. In Senapur, the proportion of people reported to be illiterate came down considerably, whereas in Seemli, this proportion marginally increased. Pandari made healthy progress in reducing the share of the illiterate population, but Gohanakallan's report card was not as impressive. The villages might have performed differently in tackling illiteracy, but when it came to raising the share of the secondary educated population, all of them accomplished average performance. Similarly, there has been a gradual improvement in the proportion of the population who have attained graduation in all four villages.

The data on the activity status of the household members revealed that the average share of the population of all four villages reporting engagement in household activities rose by ten percentage point. This rise was at the cost of allocating work, students, and the child population.

While changing the population's age structure with a rising working-age cohort has been an unmistakable sign of the development-conducive demographic pattern, the expansion of education was also discernible in reducing illiteracy across villages. Both these processes appeared to have gotten along with each other pretty well. Previous studies have also emphasised the role of education, particularly women's education, in the fall in fertility rates. But the bonhomie between the two did not last beyond basic education. As one ascended the educational achievement ladder, rural India looked like it was dragging its feet. While the educational aspirations of the people have surpassed literacy and primary levels, access to secondary and higher education is still confined to a select few.

The unforgivable failure of the Indian state in extending quality social development services to the rural poor has distorted the form and content of the transformation. When the state leaves workers to the vagaries of market forces, it tilts the balance of power in favour of the urban employers, who may exploit unskilled rural labourers sans bargaining power to the maximum without a shred of accountability. They could escape social security responsibility, knowing very well that hordes of rural migrants would continue to join the burgeoning ranks of the expendable labour force.

The failure of the state to ensure the social development of working people brought back caste, gender, and land into play. A universal and quality education and health system might, to an extent, neutralise the role of social structure as a mediator of mobility. Conversely, the failure of public policy to act as an antidote against entrenched caste and gender regimes and privileges drawn from legacy landownership would often allow these sources of power to come back with a vengeance.

Inequality in Asset Ownership

As an obvious corollary of socially mediated mobility, rural societies have witnessed a rise in inequalities on several parameters. Indian villages were always marked by inequalities along gender, caste, and class lines. But the process of transformation appeared to have accentuated differentiation among rural households. However, this differentiation phase seemed to have been triggered by rural society's closer integration with the market economy, especially the engagement of workers in non-agricultural labour and salaried jobs, instead of earlier forms of differentiation located in an agrarian structure.

But it would be erroneous to assume that both these processes are autonomous of each other. On the contrary, older forms of differentiation provided the basis for the capabilities crucial for newer forms of mobility and consequent differentiation. This was evident in horizontal mobility from one sector to another in most cases, rather than vertical movement.

The relative value of household assets was considered one of the critical indicators of differentiation. The data on the relative distribution of durable household assets revealed that, on average, a non-SC–OBC–ST household's value was more than four times higher than the value of an SC household and two and a half times that of an OBC household. This gap was three times in the case of SCs and two times in the case of OBCs in 2013.

Similarly, the distribution of financial assets among social groups changed significantly from 2013 to 2020, further widening existing massive inequality. Whereas there was little change in the relative status of OBCs, non-SC–ST–OBCs gained in the process at the cost of SCs.

Further, the data on agricultural assets also revealed a changing picture. The number of tractors and trolleys has come down in two villages – Senapur and Pandari. In contrast, Seemli has seen substantial growth, and Gohanakallan witnessed marginal growth in the number of tractors and trolleys. The number of power tillers was also increasing in all the villages. But the number of thrashing machines was showing a declining trend. The number of pump sets and sprayers was also decreasing in some villages while witnessing a marginal increase in others.

This was quite an uneven development from the perspective of transformation that was premised on a steady increase in the deployment of modern implements. Technology was no doubt considered a crucial driver of the process of transformation. But data collected from both surveys hardly supported this linear hypothesis. On the contrary, qualitative data collected from the field suggested that the rural areas of the State have been witnessing the emergence of an agricultural implements rental market. With the subdivision of landholdings and increasing fixed costs of assets, more and more farmers preferred to rent them instead of acquiring them. A connected explanation could be that with decreasing profitability, it was fast becoming unviable for farmers to possess agricultural implements that needed hefty investments. Moreover, given the de-prioritisation of agriculture in rural areas, it was considered unnecessary to maintain expensive assets.

Occupational Mobility

Previous studies have pointed out a steep fall in cultivation as the primary occupation for a vast number of workers in rural areas. However, the data obtained at two points presented a more complicated picture. Senapur was the only village where a visible drop in working-age persons reporting cultivation as their primary occupation was noticed during the reference period. In Pandari, it was also nearly at the same level for both points. Seemli recorded a slight increase, while Gohanakallan saw a visible jump in reporting cultivation as a primary occupation. When taking all four villages together, the average share of cultivators remained almost stagnant in both surveys. A quick jump in the percentage of cultivation in two villages might be a reflection of the Covid lockdown-led return of migrants who had nothing else to fall back upon except tiny family farms. The same reason partly explains the surge in reporting of cultivation as a secondary occupation.

The other aspect of the phenomenon of relatively higher reporting of cultivation as a secondary occupation could be that workers were not quitting cultivation but diversifying while keeping a toehold on it. The sharp increase in the share of cultivation in the workforce as a secondary occupation was witnessed in all the villages except Gohanakallan. The magnitude of change was substantial in Senapur and Seemli, and considerable in Pandari. A similar phenomenon was marked in Gohanakallan in the first survey when a low share of cultivation as a primary occupation coincided with very high reporting of the sector as a secondary occupation. But overall, cultivation kept its share of the workforce occupied during the reference period. Possibly, its share had already fallen significantly before 2013, and a further fall would require a break in the economic or technological sphere.

While the share of cultivation among workers had already seen a dramatic transformation in the decade leading to the first survey in 2013, animal farming witnessed equally upsetting changes during the intervening period of the two surveys. Its share as a primary occupation among the workers fell sharply during the seven years between the surveys. Earlier, cultivation and animal farming went hand in hand, but while in the contemporary scene, cultivation has been losing its primary occupational status, while animal farming has almost lost its secondary occupational status.

Unlike cultivation and animal farming, the category that has yet to find adequate attention in contemporary literature is the class of agricultural labour. Political economy works in India have mainly deployed the manual labour category, considering that this labour class was engaged in farming and non-agricultural activities. This trend had further accentuated at an incredible pace when rural manual labour spent a tiny part of their time in agriculture and were engaged for a much-extended period in non-agricultural activities throughout the year. Their income also mainly came from outside agriculture. Considering both time disposition and income criteria, it has become increasingly challenging to designate rural manual labour as agricultural labour.

From the two-point survey data, it came out quite clearly that except in Senapur, all other villages already had a meagre share of this category, averaging much below one-tenth of all workers. This figure further fell in 2020. Senapur experienced a fall of a larger magnitude, and the share of agricultural labour among all workers fell by approximately two percentage points in all other villages. The persistence of a group of workers dependent on agrarian labour in Senapur could be understood in the context of a very high proportion of the Dalit population and an alarmingly great extent of landlessness.

Further, the data revealed another interesting phenomenon: relatively high reporting of agricultural labour as a secondary occupation. The steep fall in reporting of agrarian labour as a primary occupation swelled the reporting of it as a secondary occupation in Senapur. Perhaps a segment of workers had dropped agricultural labour as their primary occupation but retained it as their secondary occupation. In the other three villages, the share of workers involved in agricultural labour as a secondary occupation decreased significantly.

Another interesting aspect shown by the data was related to combinations of occupations. Cultivation and animal farming have been a stable combination of professions. But during the intervening period between 2013 and 2020, there was a huge drop in the share of those who were occupied in animal farming as a secondary occupation alongside cultivation as their primary occupation. While quitting animal farming, these workers shifted to non-agricultural labour. But at the same time, for those primarily occupied in animal farming, their participation in cultivation as a secondary occupation increased at the cost of involvement with agricultural labour. As a result, a larger share of farm labourers involved themselves in cultivation in 2020 compared to 2013, while their participation in animal farming appeared to have decreased. Similarly, an overwhelmingly high share of all those who chose non-agricultural labour as their primary occupation was involved in cultivation, almost leaving agricultural labour and animal farming, activities in which a sizable number of them were occupied in 2013.

This data indicated the emergence of a new set of combinations of occupations in place of old ones, along with a hierarchy of professions. The combination of non-agricultural labour and cultivation has replaced traditional cultivation-animal farming. Agricultural labourers and animal farmers have been quitting each other's work in favour of cultivation as a secondary occupation. Animal farming and agricultural labour have been abandoned by primarily non-agricultural labourers preferring cultivation over these occupations.

This data revealed that agricultural labour was at the bottom of the hierarchy of occupations. At the same time, animal farming was regarded just above it but below cultivation and non-agricultural work.

Gendered Occupational Structure

The occupational structure of the villages was essentially gendered. Male and female workers had different patterns of quitting old jobs and opting for new ones. Male workers appeared to have withdrawn from cultivation, animal farming, and agricultural labour and shifted towards non-agricultural labour. The share of male workers in cultivation, animal farming, and agricultural labour fell substantially during the intervening period between 2013 and 2020 in Senapur. They appeared to have shifted towards non-agricultural labour and salaried jobs, swelling both categories at a rapid speed.

A similar trend was evident in Seemli, located in another region of the state. In this village, a steep fall in cultivation, animal farming, and agricultural labour for males of working age was accompanied by a leap in the share of the category of casual non-agricultural labour. Salaried employment was already relatively high in 2013.

Pandari had a similar story, where a perceptible fall in the share of male persons occupied in cultivation and the almost vanishing categories of animal farming and agricultural labour resulted in a further surge in non-agricultural labour, crossing the majority mark by a wide margin in 2020. Moreover, the share of salaried employment among males of working age in the village has also doubled during the reference period.

While the share of casual non-agricultural labour in Gohanakallan remained stagnant in both surveys, a section of male workers appeared to have withdrawn from animal farming, and a group shifted to salaried employment.

Contrary to the pattern of males withdrawing from cultivation, the share of females of working age engaged in cultivation has seen a massive increase during the seven years between 2013 and 2020 in all the villages. Overall, the share of women engaged in cultivation has doubled during the reference period. Another interesting observation from the two data points data shows that, taking all four villages together, average female participation in animal farming has decreased drastically. While the share of males in agricultural labour has decreased significantly, female participation in this occupation has increased considerably across villages.

Female participation in casual non-agricultural labour has also come down, which again diverged from the occupational pattern of males. Regarding the average of all four villages, this share dropped significantly from 2020 to 2013.

It is only the category of salaried employment where both males and females have registered an increase during the reference period. The average share of salaried jobs for working-age females has more than doubled from a low base at the initial point.

It appeared that while males were moving to non-agricultural occupations, both casual and salaried, females were filling the spaces they vacated in cultivation and agricultural labour. Reading the data without referring to strengthening gender segregation would be erroneous. The occupational categories were increasingly becoming marked by workers of one particular sex.

The peculiarity of the ensuing transformation was evident in the emergence of increasingly gender-segregated labour. All those who hoped that the penetration of capital would demolish labour relations premised on pre-existing equations would be thoroughly disappointed to realise that the dynamic of change resulted in more gender-segregated occupation patterns. This dynamic could not be understood at an aggregate level. It is constitutive of at least five interrelated processes. First, it was *exclusive* to the workers of low-skill endowments, *displacing* micro-enterprises, *appropriating* basic economic activities, *constricting* the mobility of excluded women to new technology, and *burdening* women with the cost of family mobility.

However, one might argue that the first three were the "natural" progression of the economy during the transformation, and these processes impacted all workers. The exclusion of low-skilled workers was considered a historical process, a pass-through to a high-productivity economy. Most of the world's developed economies have transited through this route. The fallacy of this argument lies in overlooking that the market and the state in those countries were trying to skill and absorb the workers in the new technological environment. But the capital in third-world countries with open unemployment rarely bothered about it. It simply precluded them. In place of low skills, it discarded the worker.

The other site of displacement of women was alongside micro-enterprises. The flooding of rural markets with industrial goods and the consequent displacement of enterprises in the non-farm sector that primarily produced for or traded in the rural market has a cascading impact on women, who formed the bulk of workers engaged in these economic units. Their plight was further accentuated by the appropriation of women-preponderant economic activities by relatively large enterprises, dislocating and rendering them inaccessible to female workers.

A counterargument could be that the process of displacement and appropriation was not specifically targeting women. It only weeded out inefficiencies from the system, and women, at best, were collateral damage. It is true that it was not just a question of women's livelihood, but an all-out war on the productive capacities of the rural. However, this study has reiterated that any phenomenon impacts people as per their social location. The gender-segregated occupational pattern was not about "difference", with men and women taking up jobs of their "choice", but a clear attempt to consign female workers to low-productivity economic activities while males move to higher-paid employment. Women paid the cost of family mobility by shouldering the responsibility of farms and animals during the transition.

The other question was no less important. Historically, villages evolved as a habitat and an economic unit. These two aspects, which have been intertwined through the centuries despite drastic changes in the characteristics of economic activities, have begun to separate, indicating a fundamental shift in the way the rural was seen. Economic activities, except for cultivation and allied ones, have been taken over by industrial enterprises. Several activities within the broad rubric of agriculture have also been subsumed by the larger market economy through the penetration of industrially produced inputs and farm implements. In the process of closer integration of the rural economy with the broader economy, a gradual improvement in the socioeconomic conditions of the social groups might have been observed, but it has accentuated relative deprivation.

This shift also leads to higher instabilities. Though agriculture was not immune to volatilities, the new crisis emanates from a complete absence of social security in an alien locale, as was evident during the Covid lockdown in the entire country, triggering a mass exodus of barefooted migrant labourers under the scorching sun of summer 2020. The separation between habitat and economy also meant increasing the distance of the workers from community-based security mechanisms, which didn't offer anything in lieu. This in no way romanticises the idea of the village community since social bonds have been weakening as part of the broader transformation and subsequent commodification. Another source of heightened vulnerabilities was that the rural underbelly came to occupy the same location in an urban setting.

The Migrant Labour

The retarded transformation reflected the incapacity and indifference of the urban economy to absorb the rural migrants. The rural poor in the urban economy found a place in the burgeoning dispensable labour force that could be offloaded at will. The study emphasises that these extremely insecure working conditions were the root of the truncated rural transformation that never allowed workers to "quit" agriculture. Instead, they maintained a toehold in cultivation and animal farming as shock absorbers. This dependence on agriculture persisted despite a higher share of income from non-agricultural sources. The betrayal of the Indian state in providing education and health to rural workers devastated their bargaining power in the urban labour market.

The carrying capacity of the agrarian economy might have saturated, but it played a central role in augmenting income; however, it was insufficient for transforming the rural economy. The transformative push came from the ploughing back of the urban income by migrants. The integration of the rural economy with the broader market economy came at the expense of the former's productive capacities. The inferior other of the urban economy should be content consuming industrial products and reproducing labourers at zero cost to the capitalist economy. The increasing separation between the site of production and reproduction has been the hallmark of contemporary rural transformation. The peculiarity of the Indian variety of transformation was the imprints of caste and gender on the dynamics of change.

Several macro studies have also attested that the truncated structural transformation in India pushed workers from agriculture but did not provide avenues in the urban economy for their absorption. These investigations often underlined the rather muted growth of the labour-intensive manufacturing sector and, consequently, dismal performance in employment creation (Binswanger-Mkhize, 2013). The points made by these studies are well taken, but the findings of our study diverge in two important ways.

Though it is a well-recorded fact that the share of manufacturing in India has been much lower than in many other developing economies despite the country's consumer demand increasing relatively faster. The point of contention, however, was that whatever manufacturing development took place offered extremely insecure jobs to the migrants. When Binswanger-Mkhize (2013) holds "restrictive labour laws" responsible for the stunted growth of manufacturing, they overlook that even a minimalist reading of labour laws would not allow workers to be kept in a precarious condition. Both manufacturing and construction have shown the highest disregard for laws, scuttling even the remotest chances of workers staking a claim.

Added to it, a large number of rural migrants were engaged in own-account self-employment in the service sector. Domestic workers, predominantly females, lack any formal employment contract, leading to insecure jobs characterised by low wages, long working hours, and the absence of social protection. The emergence of the gig economy has swelled this number to mind-boggling proportions. Denied recognition as workers, they were bereft of any job security, social welfare, legal protections, and regular income. A large group of tiny traders and vendors constituted another section of rural migrants who, while working full-time, did not feel absorbed in the sense of ability to claim the urban economy. They might eke out a living here but would remain peripheral, disposable at will.

While Binswanger-Mkhize (2013) shares this study's findings that urban spillover was a crucial determinant of the characteristics of contemporary rural transformation, the analytical failure to link urban informality with serious consequences for truncated rural transformation led to an altogether different policy prescription. Another policy distortion appears in the form of confusing "formal" with bringing enterprises under a taxation regime and linking them with financial transaction interfaces without referring to the status of labour. If an enterprise attains formal recognition, but the working conditions of the labour remain insecure and undocumented, it would have equally damaging consequences for the dynamics of rural transformation.

This analysis necessitates a policy shift towards rural people rather than rural areas. If a policy architecture has to facilitate transformation, the twin planks of supporting rural development and rural workers would have to be attempted. The policy should be oriented towards the agrarian economy that involves many people and better working conditions for rural migrants who are vehicles of new impetus to the change dynamics. Knowing well that without equipping the rural workers to stake a claim in the urban economy, the rural transformation would persist in a distorted form, the policy aims to ensure the social development of rural people by providing them with quality education, health, housing, and nutrition. The policy architecture would also ensure the political and labour rights of the migrants in urban spaces.

Land Relations

The two-point data collected by the First and Second Rural Transformation Surveys, conducted in 2013 and 2020, revealed an alarmingly high level of landlessness among the rural populace, especially Dalits. Official statistics did not usually display the extent of landlessness due to faulty conceptualisation that considered any land owned by a particular household, including homestead plots. The rural transformation report based its assessment of landlessness only on ownership of agricultural land.

The First Rural Transformation Survey reported a high extent of landlessness in all four studied villages. The highest share of landless households was estimated in Seemli, followed by Senapur, Gohanakallan, and Pandari. Surprisingly, the huge extent of landlessness did not see much regional variation, but, as we will see in further discussion, the landless group's social composition did vary a lot.

The Second Rural Transformation Survey, 2020 revealed that despite a slight fall in the extent, absolute landlessness in terms of nil agricultural land has persisted at a very high level in all studied villages. In Senapur, more than two-thirds of the households were estimated to be landless. Close to half of the households were landless in Seemli. At the same time, landlessness persisted among one-fourth of the households in Pandari. In Gohanakallan, the extent of landlessness slightly increased beyond the magnitude recorded in 2013. While the share of sub-half-acre landholdings has varied across villages and has no prevailing pattern in all four villages, in Senapur and Gohanakallan, their share came down slightly, while in the remaining two villages, it increased moderately.

One point that came up quite clearly was that, historically, each village's landownership structure was unique and evolving differently. The average (inclusive of all four villages) share of holdings in the range of 0.6–1 acre might have shown some stability during the reference period, but wide variations existed across villages. For example, in Senapur and Seemli, it went up marginally, while in Pandari, it came down slightly, and in Gohanakallan, it remained almost stagnant.

Similarly, the average holdings of 1–2.5 acres range in all four villages has slightly gone up, but changes varied across villages at the village level. In Senapur, the share of holdings in this class recorded an increase, while Seemli witnessed a fall. Both of the remaining villages retained a share of this class of holdings in 2020, similar to 2013.

In the marginal category, the share of holdings in Senapur increased slightly, but the comparable figure escalated significantly for Seemli. Pandari was the only village where the share of marginal holdings remained steady on both points, while in Gohanakallan, it somewhat came down. It must be noted that at the cumulative level, the average share of marginal holdings increased by two percentage points from 52.5 per cent in 2013 to 54.4 per cent in 2020.

The above data underlined that averages relied upon in large data sets hide rather than reveal the rural situation. Each village on its own was undergoing a massive transformation that official statistics failed to capture.

On average, smallholdings have slightly increased. The share of smallholdings remained the same at both points in Senapur, while Seemli, Pandari, and Gohanakallan witnessed a minor increase. The share of medium holdings was estimated to be similar in 2013 and 2020. On this account, Senapur and Gohanakallan witnessed slight upward and downward movements, respectively. At the same time, the share of medium holdings in Seemli and Pandari increased faintly. Large holdings experienced a slight increase in their share in 2020 compared to 2013. In Senapur and Gohanakallan, no large holdings were recorded in both 2013 and 2020. The percentage of this class of holdings was stagnant in Seemli at both points, while in Pandari, a slight increase was estimated.

The First and the Second Rural Transformation Surveys estimated that the incidence of landlessness among SCs of Senapur came down by around five percentage points in 2020 compared to 2013, but still, majority of SC households were estimated to be landless. On the other hand, the incidence of landlessness among OBCs of Senapur remained well over one-third at both points. Both SC and OBC groups experienced some mobility within the marginal category. However, despite a slightly changed landownership pattern, the acute marginalisation of enormous landlessness and concentration in lower landholding categories persisted during the reference period. This is confirmed by the fact that the composition of the landless category remained almost intact at both points. This category consisted of three-fourths of SC households in 2020, with an insignificant fall compared to 2013, while OBCs constituted close to one-fourth and there was almost nil representation of non-OBC–SC–ST.

Seemli village in the Muzaffarnagar district of Western Uttar Pradesh was no exception to the general pattern of high landlessness among SCs. Close to half of the households belonging to this social group did not own any agricultural land. Another one-fifth of households owned less than one acre of agricultural land. Close to two-thirds of OBC households were reported to be landless in 2013. The extent of landlessness was greater among OBCs than SCs because the former was composed mainly of Muslims, revealing much higher landlessness. Two-thirds of this Western UP village's landless category was comprised of OBC households. However, it was a different scenario compared to other villages. Backward Muslims emerged as the most marginalised group in terms of landownership. SCs constituted an additional one-third of the category. In 2020, the incidence of landlessness among SCs had come down significantly. But landlessness among OBCs persisted at the same level in 2020. Consequently, the composition of the landless category in Seemli changed in 2020 compared to 2013, with an increase in the share of OBC households by ten percentage points and a comparable fall in the share of SCs. During the same period, the share of SCs in the sub-half-acre category increased, accompanied by an equal fall for OBCs. The most dramatic changes occurred in the half- to one-acre range, where SCs increased their share significantly at the expense of OBCs. But holdings in the range of 1–2.5 acres experienced little difference. SCs increased their stake in the smallholding category moderately but lost almost an equal quantum in the medium category.

In Pandari village of the Chitrakoot district, around two-fifths of SC households were landless in terms of agricultural landownership. The comparable proportions for OBC and non-SC–OBC–ST were one-fifth and one-tenth, respectively. Quite clearly, a larger share of the non-SC–OBC–ST group found a place in the higher landholding category than the two other groups. The data on the composition of the land class categories revealed the marginalisation of SCs and OBCs. The landless category comprised about two-thirds of SC households, followed by one-third of OBCs and a few non-SC–OBC–ST households. Against this, the medium category comprised over two-fifths of non-SC–OBC–ST households, one-third of SC households, and one-fifth of OBC households. Half of the smallholding category was constituted by the OBC households, while non-SC–OBC–ST and SCs contributed one-fourth each. The stark inequality in the ownership of agricultural land was most visible in this village. In 2020, the absolute landlessness among SCs and OBCs had come down slightly in the Bundelkhand village but increased marginally in the non-OBC–SC–ST group. Against this, the non-OBC–SC–ST and OBC groups had increased their share in the medium category significantly, at the expense of the SC group. In the smallholding category, SCs and the non-OBC–SC–ST group witnessed their share shrink marginally and went to OBCs. There was no large holding in the village in 2013, but seven large holdings were reported in 2020, and non-OBC–SC–ST groups owned six.

Coming to Central UP, about two-fifths of SC households in Gohanakallan were landless in 2013 due to owning nil agricultural land. The comparable shares for OBC and non-OBC–SC–ST groups were about one-fifth and one-third, respectively. Almost negligible SCs, a few OBCs, and over one-fourth of non-OBC–SC–ST households were smallholders. A fraction of SC and OBC households also held medium-category holdings. As with all other villages, the landless category was primarily constituted by SC households, followed by OBC and the non-OBC–SC–ST group. The category of smallholdings comprised around two-thirds of OBC households, followed by approximately one-third of non-OBC–SC–ST households and one-tenth of SC households. The high share of landless households among SCs persisted at the same level in 2020 as in 2013 in Lucknow village. The incidence of landlessness in terms of ownership of agricultural land continued to be at the same level among OBCs and the non-OBC–SC–ST group. The proportion of OBC and SC households in the smallholding category remained at the same level. However, the share of the non-OBC–SC–ST group in the smallholding category went up considerably.

The increasing area under tenancy was an unmistakable feature of contemporary rural transformation in north India. This might not be surprising for many, but market dynamics and social inequalities reflected upon it, giving it a unique Southern shade distinct from Western historical experiences. This is one of the pathways that flowed from and reproduced pre-existing inequalities among social groups. Agricultural land tenancy is a symptom of anomalies in the landownership structure. The persisting incongruencies in landownership could be corrected by state-mediated transfers, often known as land ceiling measures, or market-mediated sale transfers. North India has been known to have skipped both, boosting the tenancy market.

The longitudinal data confirms these two evident patterns. First, the area under tenancy increased, reflected in the widening difference between the owned and operated areas. While there has been little change in the landownership patterns, the operational holding arrangement was transforming rapidly. Second, the non-SC–OBC–ST group has been increasingly leasing out land, while SCs increased their leased area. The survey data revealed that OBC operated the maximum area in all the villages except Senapur. In all four villages taken together, this group possessed half of the area, followed by one-fourth of SCs and one-fifth of the non-OBC–SC–ST group.

Given the landlessness among SCs, it was not surprising that they emerged as the net lessee, and the non-OBC–SC–ST group continued to be the net lessor with their persistent hold on land. SC households of Senapur increased the area under their possession by eight percentage points in 2020 compared to 2013. In Pandari and Gohanakallan, too, this group gained seven percentage points in each village during the reference period. Only in Seemli village, the SC group lost four percentage points share in the operated area. The OBC group's share in the operated area marginally came down in three villages. In Pandari, their share remained steady on both points. The non-OBC–SC–ST group lost a considerable share in Senapur, Pandari, and Gohanakallan. Undoubtedly, the land was flowing from the UC households to the SC tenants.

At an aggregate level, the share of leased-in land was over 15 per cent of the total operated area. It was 22 per cent in Pandari, 19 per cent in Senapur, and 12 per cent in Gohanakallan. In 2020, the share of leased land in the total operated area had increased by five percentage points at the aggregate level. This was due to an increase in this proportion across villages. For example, in Senapur, the share of the leased-in area in the total operated area increased by four percentage points, Pandari witnessed a surge of five percentage points, and Gohanakallan recorded an upward movement in the area under tenancy by six percentage points. On average, the SC group reported an increase in the share of tenanted land in the total operated area by 11 percentage points. The increase was two percentage points for OBC. The only non-OBC–SC–ST group recorded a drop in the area under the tenancy.

Uttar Pradesh had a high share of irrigated area since the definition of "irrigated land" included private and public sources. The same was evident in studied villages too. Except for Pandari, all other villages had a relatively high share of the area under irrigation from either public or private sources. SCs had a lower percentage of irrigated land than other social groups. At an aggregate level, 95 per cent of the total operated area was under irrigation. In 2020, the share of irrigated land to the total operated area reported in all four villages increased by two percentage points compared to 2013. SCs gained four percentage points, followed by non-SC–OBC–ST three and OBC one percentage point.

The sluggish land market was another noteworthy feature of contemporary rural transformation. Except in the villages near urban areas, the frequency of land sales has been very low. This again goes against the received wisdom. During the seven years between the first survey in 2013 and the second survey in 2020, only 77 households in all four villages purchased land. These households constituted about 4 per cent of the total households. The highest number of households (29) purchased land in Gohanakallan, followed by Seemli (23), Senapur (16), and Pandari (9).

In all four villages, 72 households purchased agricultural land, while 15 bought homestead land. In Gohanakallan, 27 households purchased farmland, and seven bought homestead land. Comparable figures for Seemli were 20 and 7, respectively. Sixteen households of Senapur purchased agricultural land, and one household bought a homestead plot. In Pandari, nine households bought agricultural land, while no household purchased homestead land during the reference period. A total of 94 households sold land during the reference period. These households were about 5 per cent of all households in the four villages. The highest number of seller households was in Gohanakallan at 49, constituting 9 per cent of surveyed households in the village, followed by 33 in Seemli (9 per cent), seven in Pandari (2 per cent), and 5 in Senapur (0.8 per cent).

In terms of area bought and sold during the reference period, Gohanakallan topped the list with a sold area of 56 acres and a bought area of 22.78 acres. Seemli recorded a purchased area of 21.4 acres and a sold area of 17.7 acres. The comparable figures for Senapur were 14.1 acres and 1.3 acres. Pandari had the smallest area bought and sold. All in all, it emerged from the data that the land market was most active in Gohanakallan and Seemli, both in terms of the number of transactions and the area transferred. Since the entire area bought and sold was not equal to each other, it was evident that the parties outside the villages were also involved in land transactions. In Senapur and Gohanakallan, marriage expenses emerged as the foremost reason for selling land, while in Pandari, it was medical expenses. Seemli households had reported "other" reasons for the sale of land.

The high share of tenant farmers leaves no possibility of conducting agricultural policies while maintaining the iniquitous landownership structure intact. Though contemporary tenancy is quite different from its earlier incarnation, it surely squeezes out the surplus in the form of land rent, leaving the peasantry cash-strapped and unable to make productivity-enhancing investments. The persistence of a high level of tenancy is undeniably a symptom of anomalies in the landownership structure that continues to be a major obstacle to productivity improvement.

As in other problems, market fundamentalists quickly advised that leaving land relations to the market was the only way out of the impasse. The World Bank's report, India – Land Policies for Growth and Poverty Reduction (2007), exemplifies this tendency. The Report recommended regularisation of tenancy where it was illegal and leaving terms of the lease to market forces where it was regulated. It also advocated "liberalising" the land sale market by diluting provisions on land use changes from agricultural to non-agricultural and facilitating direct negotiation between farmers and buyers. The report has influenced public policy significantly in several Indian states that have diluted protective provisions for marginalised groups.

The World Bank report overlooks that agricultural land tenancy might not be a semi-feudal phenomenon, but its close interface with rural power structures persists. Whether sharecropping, fixed rent, or reverse tenancy in which rich farmers lease land from small ones, the powerful party dictated the lease terms. There has been sufficient evidence that the caste structure is rooted in land relations.

The objective of public policy should be to give ownership to the poor, strengthening them economically and socially. Giving them loans to buy land from the market would only spike the land prices, further bolstering the position of the landed classes. The immediate task should be to delink government schemes on crop insurance, conditional cash transfers, and compensation for natural calamities from landownership and extend the benefits to tenant farmers. Moreover, the misnomer that the scope of ceiling surplus distribution has been exhausted perseveres despite the evidence to the contrary. There is no shortcut to the complexities of land relations. Imposing a ceiling on landownership and redistributing the acquired land is an effective policy instrument to address the incongruities in the landownership structure.

These people, a high share of rural workers and an increasing absolute number alongside a rising population, continue to persist on the limited availability of agricultural land. One implication of the increasing pressure on land was the almost vanishing of village commons and waterbodies, consequently impacting local ecology and small farms. The oft-repeated "encroachment" issue was just one aspect of the disastrous alteration of the rural ecosystem; the state policies have also aided this process. For instance, it was instructive to see how land reform policies in India transferred the burden of landlessness to rural ecology. When the state initiative failed to obtain ceiling-surplus land from the landed class because of deliberate loopholes in the law and botched implementation due to the complicity of the establishment with dominant castes, the acute problem of landlessness was sought to be addressed by distributing village common lands. The distribution of common lands traditionally set aside for being unfit for cultivation did not offer any major relief to the landless classes, particularly regarding livelihood and productive farming. But the price of defending the interests of landed classes was paid by ecological destruction.

Another obvious implication of increasing population pressure was the per capita availability of agricultural land. However, land-augmenting measures like extending irrigation infrastructure for raising crop intensity, land consolidation drives for defragmentation of land parcels, and subsiding chemical fertilisers and HYV seeds seeking higher productivity have been undertaken, but they have proved to be insufficient in raising per capita income from agriculture. Besides, it was no secret that the "green revolution" practices negatively affected fragile ecosystems, an issue that we discuss in some detail in the next section.

Agriculture

Another peculiarity of the ensuing transformation dynamics was the rising productivity and shrinking diversity of crops. The surveys revealed an alarming trend of shrinking diversity of cropping patterns during the seven years. For example, in the Kharif season, the 2020 survey recorded only four crops: paddy, pulses, corn, and vegetables, compared to six crops recorded in 2013, namely paddy, pulses, coarse grains, groundnut, and vegetables. In addition to the annual crop of sugarcane, the Rabi season had five crops in 2013, viz wheat, pulses, oilseed, coarse grains, and vegetables. It came down to four crops seven years later: wheat, pulses, oilseed, and coarse grains.

The relative preference between crops has remained mostly the same during the period. Paddy continued to command the highest area during the Kharif season in Senapur and Gohanakallan. But Pandari only had pulses, dropping coarse grains, oilseeds, and groundnut from its cropping pattern. On the other hand, the area under vegetables has increased substantially in Gohanakallan. Senapur did not cultivate coarse grains and groundnuts in 2020 but instead shifted slightly to corn. Seemli continued to cultivate sugarcane during both seasons.

The average productivity of paddy in 2020 increased by 6.14 quintals/acre compared to 2013. It was a jump of 55 per cent in seven years. This increase was highest in Gohanakallan at 6.57 quintals/acre, followed by 4.5 quintals/acre in Senapur. The average wheat productivity in the seven-year reference period increased by 2.6 quintals/acre. It was an increase of 25 per cent between 2013 and 2020. The highest rise in productivity levels was observed in Senapur at five quintals/acre, followed by 3.5 quintals/acre in Seemli, 2.07 quintals/acre in Pandari, and 1.5 quintals/acre in Gohanakallan.

Broadly, the agriculture sector in the study area, as in larger parts of the developing world, was going through a massive churning led by a mechanisation spree. The tractorisation of agriculture that began with the green revolution in the 1960s has become unstoppable with the addition of newer and heavier machines and is spreading every nook and corner of the country. This scenario might have surprised those who thought mechanisation was for the labour-scarce West while the labour-surplus Global South would follow the path of investment in productivity-enhancing inputs and irrigation infrastructure (Dorin, 2017). Remarkably, the spread of farm implements has ceased to care about the landholding size, thanks to the emergence of the rental market, where large farmers often lease out after completing their farm operations. The arrangement might not be the best suitable option for the smallholders, but it provided access to them. The demise of the ecosystem that nurtured animal-farm cohabitation has also pushed farmers' reliance on machines.

It is an obvious expectation that a mechanisation drive would "release" workers from the sector, but the family farms of the developing countries had another surprise in store for those looking forward to a predetermined path. Over two-thirds of workers remained occupied in agriculture as cultivators, animal farmers, and agricultural labour in either primary or secondary status. The displacement of manual labour from farm operations has been a continuous process, the latest addition being harvesting with the arrival of the harvester-combine. The machines already overtook several post-harvesting operations. However, manual workers continued their hold over labour-intensive activities like paddy transplanting. It was intriguing that reporting of agricultural occupation did not fall commensurate with the deployment of machines. In the absence of alternatives, workers appeared to be attached to the sector despite shrinking opportunities and falling income. It was not rare to find middle-aged women from poor households who barely got a few days of farm work on a paltry wage.

Alongside the atypical propensity for displacement, a major shift has taken place in the composition of the labour force occupied in the sector. While men moved on to higher productivity non-farm jobs, women filled the space in cultivation and agricultural labour. Only in animal farming, as a primary occupation, have the falling declining trends of males and females converged. The participation of social groups in cultivation appeared rather stable, but Dalits and OBCs shifted their base from animal farming and agricultural labour to casual non-farm labour. The destination for the "general category" withdrawing from agriculture was primarily salaried jobs.

With many people dependent on agriculture, the agricultural policies of developing countries, specifically India, have been found wanting for their excessive focus on production and productivity, with the complete abandonment of the labour question. A typical headline policy brief begins and ends with the production quantities without mentioning the people beneath it.

Notwithstanding some tinkering, sometimes as crisis management and other times under popular pressure, the archaic policy framework still saw agriculture as a means of feeding the population, whereas workers could only be better off after leaving agriculture. Astonishingly, the largest employer in the Global South has never been considered capable of creating quality employment opportunities. The policy orientation of looking at agriculture as a transitory location for workers has percolated to the social common sense where those who stayed in agriculture would be treated as "left behind". The falling income due to market vagaries has only buttressed this perception. The policies needed a reorientation towards people's livelihoods, shunning the baggage of a successive sectoral transformation.

Reorienting Agriculture Policy

Some of the recent works provide a clue to breaking the agricultural impasse. Diverging from the dominant industrial tendency of conquering nature, the alternative agricultural policy should strive for a harmonious relationship between humans and ecology. This would necessitate shunning the misnomer that higher productivity is synonymous with extensive usage of natural resources like water, energy-guzzling machines, and agrochemicals. The stubbornness of treating agriculture as a factory system, where similar production processes could be applied globally, will give way to an agroclimatic region-specific strategy. This could only be done by overcoming the arrogance of outrightly rejecting traditional knowledge systems of flora and fauna and, instead, attempting to combine local experiences with science. An alternative approach might not end all agricultural woes but would prepare cultivation for sudden and extreme weather conditions. A labour-intensive, pesticide and chemical fertiliser-free agriculture would be income-enhancing for farmers and healthier for consumers. This approach also calls for embedding agriculture into the secondary and tertiary sectors by offering value-added products and ecological services (Dorin, 2017).

At a more fundamental level, Patnaik and Patnaik (2016) locate the woes of petty producers in the dynamics of capitalism at the global level. They argue that temperate regions of the North could not produce all the commodities, leaving them dependent on the tropical and sub-tropical regions in the South. As a result of increasing demand for certain commodities, the risk of escalating prices posed an imminent threat to the value of money, the basis of capitalism. To thwart this occurrence, the consuming countries imposed income deflation on the producing countries, causing acute hardships for peasants and artisans. This imperialist tendency operated in the form of direct control of one country over another country during colonialism but adopted newer mechanisms in the contemporary world marked by neoliberalism. The authors of this crucial work suggested that the Global South would have to de-link their economies from the finance capital-led globalised economy and reinvent linkages between countries of the South (Patnaik and Patnaik, 2016).

In a nutshell, agricultural policies of developing countries would have to be reoriented towards people's livelihood (Vijayabaskar et al., 2018). Once the labour question occupies the centrality it deserves, a reversal of policy direction pushing workers from agriculture destined to nowhere would be a logical first step. Even those who swear by the historical experiences of the industrialised world admit that there aren't enough jobs in industries for the humongous number of people employed in agriculture. The availability of colonies played a major role in offloading excess workers of the West. Not all the workers in eighteenth-century England were absorbed in emerging industries (Patnaik and Patnaik, 2016). In the populated countries of the South, income-enhancing measures in the agricultural sector would go a long way than attempting to imitate the Western-style transformation. This would entail intervention in agricultural marketing, rural credit markets, risk management, climate compatibility, and land relations.

Agricultural Commodity Markets

The policy framework on agricultural markets in developing countries generally suffers from multiple misconceptions. The framework has often been caught in a control vs open dilemma, having a close interface with another fallacious understanding of assuming agricultural markets as a homogenous construct. The actual markets were constituted by a series of agents, with petty commodity producers and small traders forming the bottom while national and international agri-businesses were on the top, not to mention finance capital playing an influential role. Moreover, markets performed multiple tasks of "efficiency, extraction and exploitation" against the mainstream notion of "efficient allocation of resources". Following Polanyi (1944) that markets were embedded in the economy and the economy in social relations, the agricultural markets manipulated primordial institutions to their advantage, debunking any possibility of autonomous markets (Jan & Harriss-White, 2012, p. 46).

Market functioning needed state support in the form of regulations, enforcement of contracts, maintenance of law and order, and political stability, to say the least. States often came to the rescue of markets in cases of crisis as well. In this discourse, the implicit meaning of the "free market" was strengthening the position of the powerful party and leaving the weaker party unprotected (Jan & Harriss-White, 2012).

This tendency could be seen in the three Farm Laws (now withdrawn) enacted in India in 2020. The first law, the Farmers' Produce, Trade, and Commerce (Promotion and Facilitation) Act 2020, sought to remove all "hurdles" in free trade outside the market yards regulated by the Agricultural Produce Marketing Act (APMC), consequently creating two markets. Concerns about the absence of any redressal mechanism on price and non-price issues in the open market were raised. The second act, the Farmers (Empowerment and Protection) Agreement of Price Assurance and Farm Services Act 2020, sought to promote and institutionalise contract farming. The Essential Commodities Act 2020 delisted several agricultural commodities under a protective umbrella against market manipulations by hoardings. In a nutshell, all the laws read together were aimed at withdrawing state protection from farmers, opening the floodgates of the market and, more importantly, bringing face-to-face unequal parties – the farmers and agribusinesses in an unregulated environment (Jha, 2022). Once implemented, these laws would have had adverse consequences for farmers and the food security of the country, but under popular pressure, the government had to withdraw them.

These laws were examples of how states in the Global South frequently look for easy answers to the complex problems facing agricultural marketing by handing over everything to capitalist markets. In developing countries where petty producers and small traders, though in a relationship of tension and cooperation, provide a livelihood to millions, the policy architecture should prioritise protecting the interests of these groups (Jan & Harriss-White, 2012).

The cooperatives and farmer-producer organisations (FPOs) could be the way out. The cooperative movement received uneven success across sectors and states in India. Western and Southern states like Gujarat, Maharashtra, Kerala, Karnataka, Tamil Nadu, and Andhra Pradesh, and left-ruled states like West Bengal have recorded relative successes in agriculture and allied sectors like dairy and fisheries. Gujarat's Amul and Maharashtra's cooperative sugar mills are considered landmarks in the cooperative movements of the country. Almost all states have cooperative credit societies and banks.

Despite these models, the cooperative movement in India did not yield the desired results in Northern states that have been home to small and marginal farmers. This is not the place to discuss challenges faced by cooperatives, but their failures were explicitly marked where caste-land inequalities were comparatively higher. However, no Indian state could boast of completely decolonising the land tenure system, paving the way for a relatively egalitarian rural society, a prerequisite for the success of cooperatives. Since the success of the cooperative was contingent upon the success of the land reforms programme and enabling an egalitarian environment, its failures could not be attributed solely to the idea of cooperation. The idea of smallholder economies pooling together to achieve the scale of the economy remains relevant in the Global South. It has acquired new significance in the contemporary situation where petty commodity producers are experiencing a squeeze from the global agro-industry.

Failure of the states in developing countries to ensure a conducive climate for the cooperative movement has led them to attempt new forms of collectivisation. The new incarnation was a producer company of a farmers' collective. However, the major difference between a farmer-producer organisation (FPO) and a cooperative was that the former operated in the market domain while the latter operated in the state domain. The tendency to look at the market as a universal remedy also helped FPOs gain currency. Though FPOs are in the experimentation stage in many countries, including India, where the National Policy for the Promotion of FPOs was adopted in 2013, they have shown encouraging results in aggregation at both ends – input and output of the farm economy. These newly set-up companies might be suffering from limited access to financial institutions, uneven distribution of gains among different classes of farmers, absence of proactive policy architecture in terms of company regulations, taxation and access to regulated markets, infrastructure, and sufficient handholding by promoter agencies (Trivedi et al., 2022), but collectivisation efforts of the smallholders should be a core constitutive element of public policy.

Rural Credit Market

The plethora of issues concerning rural credit markets was closely related to agricultural marketing. In policy discourse, the credit and commodity markets might figure separately, but these two markets were often interlocked for the millions of smallholders in developing countries. It was not uncommon to come across farming households that had taken an advance in cash or kind from a local grain dealer or input supplier on the condition that the produce would be sold to them at a predetermined price, which was significantly lower than the market price and the minimum support price (MSP) declared by the government in India. The interlocked contracts of finance and production where the interest component was generally implicit have sometimes been hailed as serving the mutual interests of both parties (Bardhan, 1983) and minimising transaction costs (Braverman and Stiglitz, 1982); in reality, they suck out a large portion of the value produced in cultivation, leaving the debtor perpetually crisis-ridden and without any surplus for investment. The "mutual interest" and "transaction cost" views failed to see the exploitative nature of these contracts in which one party was necessarily poor. They also overlooked that interlocked contracts reinforced power relations based on land, caste, and gender.

Knowing well that rescuing the agrarian economy from the clutches of informal moneylenders was crucial for the enhancement of productivity and well-being of farmers, policymakers rightly suggested measures to extend formal lending to the sector. But these policies frequently resorted to the linkage between landownership and finance, effectively excluding informal tenant farmers who constituted as much as one-fifth of all farmers in some countries. The same fallacious understanding of farmer and landowner being coterminous underlies conditional cash transfer schemes, crop insurance schemes, and natural calamities compensations, among others.

Ecological Sustainability

The fourth major site that urgently required policy attention in the countries of the Global South was the ecological sustainability of agriculture, which precisely meant saving the ecology from adverse impacts of farming and protecting cultivation from increasingly unpredictable weather conditions.

Agriculture is considered one of the major contributors of greenhouse gases in many developing countries. For instance, as per the First Biennial Update Report to the United Nations Framework Convention on Climate Change (2015), the share of agriculture was about one-fifth of the total greenhouse gas emissions in India. This figure might be revised upwards if emissions on account of the manufacturing of pesticides and fertilisers were added to it. Wet rice cultivation, chemical fertilisers, and crop residue burning were considered bigger sources of emissions. Besides the emission of greenhouse gases, excessive usage of chemicals for fertility enhancement and pest management has resulted in an ever-enlarging presence of harmful chemicals in flora, fauna, natural waterbodies, soil, and air. Disproportionate usage of chemical fertilisers has been identified as one of the contributing factors to soil degradation (Veluguri et al., 2019).

While agricultural activities adversely affect the ecology, cultivation has been increasingly prone to extreme weather events. Record-breaking heat waves, cloudbursts, floods, and droughts, sometimes two or more coinciding with each other, have devastated agriculture in various parts of the globe. The rhythm of agriculture gels with predictable and compatible weather conditions; whenever there was a variation, it was within permissible limits. Extreme weather conditions were not unknown, but their frequency and intensity have increased in the last few decades. Tripathi and Sindhi (2020, p. 5) have summed up the adverse impacts of climate change on agriculture in the following words:

> First, climate change impact in terms of temperature and rainfall is non-linear and is felt in the extreme, that is, when temperatures are higher, rainfall is substantially lower, and the number of dry days are higher than normal. Second, is that these extreme shocks have highly divergent effects between unirrigated and irrigated areas, almost twice as high in the former compared with the latter. Third, it relates to the impacts on agricultural yields. The estimate shows that the extreme temperature shock reduces yields by 4% and 4.7% for kharif and rabi, respectively, while the extreme rainfall shocks reduce yields by 12.8% and 6.7% for kharif and rabi, respectively.

Saving agriculture from the vagaries of the changing climatic conditions is easier said than done. Agrometeorological services providing advanced warnings and allowing timely planning could be helpful in the short run (Lobo et al., 2017), but it is no more than firefighting. There have been other ongoing experiments with climate-resistant seed varieties, but weather conditions ranging from extreme drought to extreme wet in a short span could present insurmountable challenges. But slowly reducing the harmful effects of industrialised agriculture, often known in developing countries as green revolution practices, could be the starting point. Some alternative agricultural practices have emerged, including organic cultivation. The policy should move towards post-green revolution agriculture marked by harmony between ecology and cultivation.

Rural Non-farm Sector (RNF)

In one more crucial way, this study's findings resonated with Binswanger-Mkhize (2013). It was about de-linking the rural non-farm sector (RNF) from agriculture. Traditionally, the fate of the rural non-farm sector was viewed as closely tied to cultivation and animal farming. But the contemporary non-farm sector was more an extension of the urban economy, almost uprooted from the agrarian economy. The sector drew resources from the urban economy and created a market for industrial products.

In Binswanger-Mkhize's data, the emergence of rural construction has been highlighted for its increased job share. However, trade, transport, and communication have also registered equally impressive growth, and despite losing share, manufacturing maintained salience until the beginning of the new millennium. The findings of our study diverged from Binswanger-Mkhize's work both spatially and longitudinally. One must admit the diversity in rural situations across India; much has changed in the last two decades.

Not discounting the importance of construction that has added new momentum to the rural economy and diversification of employment, the rural economy has emerged as a consumer of industrial products and services. The new RNF offered urban services in rural hinterlands. This is what gives a semblance of urbanity to the countryside. Importantly, it connects rural with urban, making their boundaries more fluid. This proximity to urban areas prompted Binswanger-Mkhize to conclude that non-agriculture was not necessarily urban, and the models would be compelled to accommodate the RNF. The challenge to the tendency of equating rural with agriculture and urban with non-agriculture was well taken; however, Binswanger-Mkhize, too, found this terrain quite slippery.

The problem with Binswanger-Mkhize's analysis was its straitjacket manner. The assumption that loss of share in the employment of one sector and the rise of another were mutually exclusive ran through. Most of the macro studies shared this lacuna. The longitudinal data told the other side of the story. If "quitting" agriculture was a prerequisite for transformation, political economists should prepare for a great disappointment. Dumping cultivation was not the first thing on people's minds, even while labouring to garner relatively better income from other sources. This multiplicity was not new but acquired a new meaning across sectors, locations, and seasons.

Moreover, the urban spillovers were not exclusively directed towards RNFs. The rural transformation data found that labour wages were one of the major channels for siphoning urban funds into the rural economy. The qualitative data revealed that a major portion might have been spent on purchasing goods and services, but a considerable share was also used in agriculture as working capital and for the acquisition of capital goods. The public policy on rural development must acknowledge and respond to the multiplicity of occupations and income sources.

Transformation of Social Relations

An economistic view has coloured the public policies on rural development in much of the Global South. It was in line with the rural transformation studies that have often been found neglecting the social relations that closely correspond with the economic relations. The neglect of social relations was partly due to the understanding that they would change in tandem with underlying economic relations, sometimes with a lag. They were also consigned to the cultural domain that had nothing to do with the economy.

The bigger takeaway from the present study was that social relations did not necessarily change in tandem with the economy, though both were closely related. Even when both were changing, the dynamics of change might be different, at times going in dissimilar "directions". Together social, economic, and political changes debunk the notion of a linear change that has often informed the policy formulation in developing countries. Social policies have been chasing the Western model all along, frequently overlooking the complexities of the Southern countries.

It was no more evident than in the case of women's falling labour force participation rate (LFPR) since the mid-1990s, which showed signs of recovery after 2019–2020. It has been seen as validating the U-curve hypothesis. The pattern was taken for the withdrawal of females from the labour force due to the growing prosperity of households that were no longer dependent upon women's earnings. The falling LFPR might not always be bad news. Sometimes, women withdrew from work to pursue education and re-enter as educated workers. Moreover, they could also withdraw due to social constraints (Pandey, 2023).

Since the withdrawal was attributed to "income effect" and "education effect" (Pandey, 2023), which were seen as positive impacts of the transformation, it did not yield any comprehensive policy response. Moreover, this was also viewed as validation of the new economic policies adopted since the 1990s. While the education and income effects could not be denied, the women's movement to fill the gap created by men's diversification reflected a lack of opportunities within the reach of female workers. The emerging jobs were spatially distant, or workers lacked the required skill sets.

Moreover, the occupational pattern of the studied population was not only gendered but also caste-segregated. It has been noted frequently in sociological literature that taking a cue from the landed classes' anxiety to control women's bodies and sexuality, which confined women to homes, was primarily an "upper-caste" practice. Subaltern women have been traditionally active contributors to the economy world over. The income effect subsumes this factor, known as Sanskritisation, when the lower caste people imitate upper caste cultural practices. The overall pattern of change might have been U-shaped, but it was constituted by numerous socio-political processes that should be of immense interest to public policy.

The sex-segregated occupational pattern was not a new thing, but its being the constitutive element of the dynamics of rural transformation belies hopes that gender-based inequalities would wither away with capitalist modernisation. The role of public policy acquires renewed importance when the spontaneous dynamics of a social formation fail to address certain deficits.

Property relations were one of the most crucial sites requiring state intervention in seeking gender justice. In larger swathes of the Global South, agricultural land was the most valuable property in rural areas. Its unequal distribution drew from many countries' old landownership structures and patrilineal succession legal regimes. For instance, the succession provisions in the Revenue Code of the north Indian state of Uttar Pradesh have been designed to contain agricultural land in a male-headed family. Women do have some rights. Unmarried daughters and widows found a place in the first class of heirs, but unlike male descendants, they tend to lose their rights upon a change in marital status (Trivedi, 2022). The property succession regimes in these countries needed an overhauling to be gender-neutral, allowing equal rights to all descendants. The question of gender justice in landownership acquires a renewed urgency as the rising number of women farmers would also need secured ownership rights and enabling social spaces.

Another reason the revenue laws of many developing countries needed to be overhauled was to loosen the grip of the traditional elite over agricultural land. In most of South Asia, these landowning groups comprised upper castes and some other dominant castes, excluding a large population of lower castes, particularly ex-untouchables. The disproportionate land distribution in these regions has long sustained iniquitous social and power relations, posing a serious challenge to democratising rural society. No doubt, landownership patterns, social relations, and local power structures have negotiated the contested rural spaces. Dalits in north India might have taken the route of withdrawal from the oppressive agrarian structure after the dismal performance of the state in curbing the power wielded by the landed classes, but the truncated rural transformation has kept the land question alive. Breaking down the caste–gender–land equation was the prerequisite for any conceivable notion of rural transformation.

Similarly, instead of taking women's withdrawal from paid work as a natural course of history, the policies must work towards recovering the productive capacities of rural areas. This research and many other works in the field have shown that the loss of compatible job opportunities has contributed immensely to women shifting from paid employment to unpaid household work. While women's turn to unwaged work was part of a broader process of shifting economic realities, it entails a task cut out for public policy to address the exclusive character of the emerging economy. The policy must work on two fronts to bridge the skill and spatial distance between jobs and workers.

Plenty has been written about the skill deficit among workers in developing countries that has left them "unemployable". The industry often whines about the lack of trained "manpower" that could keep pace with a fast-changing technological environment. These deficits have been found to be higher among women workers. Nothing is intrinsic about being a woman and not being trained in a particular skill. Despite complaining, the industry has contributed little to the skilling of workers in general and women in particular. The state would have to come forward to salvage the situation.

The rural might have imitated the urban in numerous ways, but it simply lacked the educational institutions imparting quality education to the poor. Government schools in north India were in shambles. Though these schools provided free education to poor students, the deficit in the number of teachers, their involvement in several government duties, and lack of infrastructure resulted in low learning levels (ASER, 2022). In the education sector, governments lacked ambition comparable to the infrastructure sector, where huge public investments have been going in the new millennium. Everybody needed roads and electricity, but the poor needed public education and health more than anything since they could not buy it from the market. And whenever they attempted to dive into the education market with meagre funds, they mostly ended up in worse private schools (Nambissan, 2012).

There was no shortcut other than strengthening the public education system. In addition to the instrumental value of preparing "job-ready" workers, public institutions comprised students of different castes and communities. Though public schools have witnessed discrimination against SC and minority students, one could only hope that interactions between students might have prevented the hardening of caste attitudes (Nambissan, 1996).

The most challenging task before public policy in the Global South is addressing the growing separation between the site of production and the site of reproduction. This can only be achieved by restoring the productive capacities of rural areas. This will bring employment opportunities closer to workers, especially women. Currently, the policy has been paralysed by two misconceptions. The first one is about a perceived connection between rurality and elementary technology. And second, a misreading of large industries and modern technology. The policy must emphasise that the non-farm sector can go beyond food-processing industries. It can produce anything conceivable on earth, not necessarily in large capitalist industries. Experiments in other Asian countries have shown that small village enterprises can also be a thriving space for modern goods.

Concluding Remarks

In the introduction of this book, six dimensions of a reimagined rural transformation were identified. They relate to the breakdown of jati and gender congruence with landed property and the commodification of land, as well as a shift from "status" to "contract" for labour. This shift reflects the freedom to choose one's employer without constraints, with wages determined by the market without discrimination. Diversification of employment outside the agriculture sector constituted a crucial element of transformation. These changes must lead to a better life in terms of social development – access to quality education, health, nutritious food, housing, drinking water, social equality, and political participation. Finally, a heightened level of investment and deployment of newer technology was considered both a driver and an important site for the study of transformation.

The data presented in the book chapters showed a more complex picture than expected. It found that some aspects of the rural economy and society were transforming faster, while others gave a robust defence against the change. Alongside, several other facets modified their form, while others retained it.

But this rather muddy description shall not lead one to miss the wood for the trees. A society under massive transformation might take various forms depending upon historical contingencies and the balance of social forces. Still, capital was essentially restructuring every single facet of rural life. No aspect of social or individual life has remained untouched by a massive wave of capitalist transformation sweeping rural India. However, villages were never isolated, and previous rural transformation surveys also captured the ongoing penetration of capital in the agriculture and non-agriculture sectors of the rural economy. The seven years after the first survey have seen further expansion of a similar tendency of integrating the rural economy with the market economy.

However, the most interesting aspect of the transformation experienced by rural areas was related to the modification of pre-existing relations rather than a complete replacement. This process gave way to a new set of relationships that were so reminiscent of older forms while exhibiting features of the new ones. The following discussion will describe the emergence of a new set of relations in different aspects of economic life.

The most interesting transformation captured by the surveys related to occupational structure when subjected to gender and caste disaggregation. Occupational diversification has been a much-talked-about subject in social sciences, but the dynamics of this transformation were rarely captured. For example, it was superficial to assert that recently, workers in the agriculture sector were moving to the non-agriculture sector. Like any other aggregate picture, this was a half-truth, if not outright misleading.

This kind of generalisation needed to be more accurate since the data collected by both surveys revealed that the diversification in India was not a break from the past. Workers were rarely quitting the agriculture sector but diversifying outside it while maintaining a toehold in it. However, in terms of both time disposition and income, the non-agriculture sector was the main occupation for the vast majority of the workers, but the agriculture sector provided them with a shock absorber. Some of the findings of Lerche, Shah, and Harriss-White (2013) were also on similar lines. The Second Rural Transformation Survey conducted in the aftermath of the countrywide lockdown in 2020 re-established the wisdom of rural workers for keeping links with the agriculture sector alive. In times of acute crisis, when all the economic activities came to a standstill and migrant workers returned home battered by long walks and extremely tiring journeys, they found something to hold on to in their tiny plots.

For some social scientists, it was surprising to see rural workers cling to agriculture that wasn't generating sufficient income, if not an outright loss-making enterprise. Perhaps their vision was blurred by the inability to see the highly insecure nature of the jobs and "self-employment" in the non-agricultural sector. But workers have always been aware of their working conditions and kept contact with agriculture as a cushion in various forms: cultivators, tenant farmers, and agricultural labour.

These peculiar characteristics of the "formal" sector essentially determined that workers did not quit agriculture to join jobs in emerging sectors of the economy. This link between industry and agriculture was crucial in understanding the nature of capitalist development in rural areas. This model of industrial development in India completely diverged from the Western experiences where the emerging industrial sector completely absorbed a large section of newly arrived workers from agrarian backgrounds. This was not happening in India.

The emerging sectors of the economy in India, whether construction, small-scale industries, or others, did not absorb workers; instead, they adopted the strategy of maximum exploitation by keeping them only partially admitted into the workforce. This arrangement allowed them to deploy these workers at will without owing any responsibility. This super-exploitation of workers might be making the Indian industry "competitive", but on the other hand, it resulted in a truncated transformation of the rural labour force. A large share of the workforce reporting agriculture as their occupation in official statistics reflected this reality.

Further, it would be erroneous to analyse this situation within the framework of the level of capital penetration. It wasn't a question of the pace of capitalist development in rural areas either. Instead, the nature of capitalist transformation in the country as a whole overwhelmed villages in a particular manner. They might have lost their status as sites of production and accumulation but continued to play the role of reproduction.

The capitalist restructuring that essentially assigned the role of workforce reproduction to rural areas while almost destroying their production capacities reflected the falling share of agriculture and allied sectors in the GDP of India, but more than that, it was reflected in the demographic transformation seeking to align with the demands of capital.

Before coming to the demography, one might ask that if extremely insecure jobs in the non-agricultural sector were so exploitative, why would rural workers flock to urban centres seeking and clinging to them? At a general level, with the agriculture sector plunging deep into crisis, especially for those with marginal holdings that cannot derive a livelihood even when the industry recorded relatively high growth, migrating outside was the only available option. Moreover, incomes from agriculture had fallen to the extent that a super-exploitative job in other sectors appeared an attractive alternative.

Alongside a disaggregated analysis would also reveal that the movement towards jobs fetching relatively higher income in the complete absence of any possibility at home might be a motivation for all groups of workers, but Dalit workers also strived for freedom from the structure of domination in rural areas, particularly if one was engaged in agriculture as an insecure tenant or agricultural labourer. Capital might not have "eradicated" *jati* from rural areas, but it provided avenues of caste anonymity outside direct domination.

But the capital–caste complex has one more dimension. It only provided opportunities for horizontal movement and restricted vertical movement. Achieved status might be conceptually distinct from ascribed status, but the capital–caste dynamics ensured that the emerging opportunity structure was not free from the baggage of *jati*. Workers in the studied villages migrated en-flock to new opportunities, but their social location reflected upon, if not determined, their destination.

Similarly, gender emerged as a strong determinant in occupational diversification. Whether within the agriculture sector or transcending it, patterns of female and male workers were quite different. Female workers appeared to fill the space vacated by male workers while migrating to newer jobs. The burden of low productivity and much lower-wage jobs fell upon the shoulders of female workers.

Capital might not be worried about labour supply in an economy with open unemployment. However, a steady stream of the working-age population was supportive of the economy, and population policies worldwide have been seeking to do just that. Of course, the state policies might not always be able to successfully coax people to change their behaviour on matters that have wider cultural and economic sources. But at the risk of oversimplification, it could be claimed that a compatible demographic transition would hasten the transformation process.

Whether it was a happy coincidence, an influence of policies, or a combination of both, all four studied villages were undergoing a demographic transition supportive of transformation. In all the villages, much like the macro picture of the state or country, the share of the working-age population (15–59) has grown in the last seven years. On the other hand, the growth rate of the people in all social and religious groups has come down, which was also reflected in the falling share of the up-to-14-years age cohort. A marginal increase in the percentage of the aged population may have concerned some, but a leap in the share of the working-age population could offset its effect.

The linkage between the demographic transition and the transformation of the rural areas was through their relationship with capital. In the case of studied villages, much like most villages, a larger share of the working-age population meant a higher number of workers shifting to the non-agricultural sector, often the capitalist sector of the economy, and bringing home income drawn from these sources. These incomes fuelled the transformation of the sectoral composition of income and labour, decreasing the importance of agriculture in the economy and equally critical cultural changes in lifestyles, consumption, and aspirations. It was becoming quite clear that rural integration triggered contemporary transformation with the market rather than through the internal dynamics of the rural economy. This was more akin to "capitalism from above" rather than "capitalism from below" (For a detailed discussion on the subject, see Lerche, 2013).

Rising inequality was another remarkable feature of the transforming rural under the impact of market integration. At the same time, other facets of social life had gradually changed in the seven years, but inequality between social groups had broadened qualitatively. Rural societies were never egalitarian, but the way the market interacted with structures of caste and gender, pre-existing inequalities were carried forward and further exuberated. The dynamics of this transformation were characterised by pre-existing structures and the market reinforcing each other's power. It was a qualitatively different phenomenon than Western theory's "replacement" view.

Equally important, if not more, a driver of the transformation was state intervention in rural areas. State-initiated policies, programmes, schemes, and laws have transformed the rural society and economy beyond recognition. However, much like the capital, the state's record was also mixed regarding success in transforming rural institutions.

The state's first and crucial failure that stood in the way of transformation from below is reflected in the survey data on landlessness in studied villages. Due to faulty conceptualisation, official statistics might have never revealed the actual extent of landlessness. But when the two-point survey data was analysed considering nil ownership of agricultural land, ranging from one-fourth to nearly half of all the households were estimated to be landless. This humongous extent of landlessness was not distributed among all social groups but was essentially concentrated among Dalits and Muslims.

The persistence of thick and unyielding congruence between caste and land that testified worst failure of the state in conceptualising and implementing land reforms was captured by the surveys in the not only concentration of landlessness among certain social groups but also hold of specific groups on land. Inter-generational subdivisions might have reduced the average size of the holding. However, the entire agricultural land was still owned by the group that held it before the abolition of intermediaries.

The state might have failed in de-colonising India's land tenure system, resulting in the stunted transformation of rural areas. But it supported the agriculture sector in raising productivity, resulting in a massive scaling-up of agricultural production. However, unintended consequences of these interventions aided the transformation of primarily labour relations in agriculture, displacing manual labour with machines for several farm operations. Often contradictory processes of mechanisation, increasing crop intensity, and changing cropping patterns resulted in the shortening of the period for which manual labour was required, but the creation of massive demand during peak periods outstripped supply. Alongside, the shift towards cash wage-based non-farm jobs ensured the breaking down of old labour relations in agriculture and a steady rise of cash-based hiring of wage labour. Ramachandran (2011) has also found the number of days of employment available to rural manual workers to be low.

These changes in labour relations in the studied villages were already evident by the time of the first survey. It was also apparent by then that tenancy had considerably persisted in all the villages. In several cases, these tenancy arrangements were nothing more than labour arrangements. An increase in the share of area under tenancy to the total operating area during the seven-year intermittent period between the two surveys told a different story than what was revealed by the transformation of labour relations.

The pattern of tenancy contracts seemed to be heading in two directions. In many cases, the persistence of sharecropping was little more than the deployment of labour while saving on managerial resources. A leap in the share of Dalits in leased-in land combined with the rising share of women workers in cultivation explained the expansion of the area under sharecropping. Mostly, women workers from low-income families who stayed back home under the umbrella of an aged man tilled tenanted land. Cash rent-based contracts were another direction in which tenancy was heading. It became more apparent in the second survey that cash was becoming the principal mode of rent payment in all the areas of the state.

Both of these tenancy contracts lead to different analytical points. In literature, the persistence of small tenancy was often seen as a sign of retarded transformation. But in studied villages, it appeared more as a consequence of developments in the market economy rather than the transformation of the agriculture sector. The steady emergence of cash rent indicated that tenancy relations were not necessarily "semi-feudal" in character. However, any rent-seeking retarded the transformation of agriculture by siphoning off investible capital, especially in cases where investment from non-agricultural sources was not available. But it was not surprising to find households that invested non-agricultural income into agriculture. Moreover, the productivity of plots under tenancy and owner-cultivation has not varied much.

In a nutshell, an increasing area under tenancy indicated anomalies in the landownership structure resulting from the failure of the state to break the land–gender–caste complex. Besides, it also revealed that landed property had yet to be wholly commoditised, which was also reflected in a low level of market transactions in land. But tenancy relations were not "semi-feudal" and moved towards cash-based contracts. Other works have also denounced the "semi-feudal" hypothesis (Shah and Harriss-White, 2011)

Another state intervention that proved crucial for transformation was the creation of critical infrastructure like roads, electricity, and transport in rural areas. Alongside the explosive expansion of telecommunication networks and the consumer goods market that rode on the back of critical infrastructure, the strict compartmentalisation between rural and urban collapsed.

However, state intervention in social development, particularly education and health, have witnessed considerable expansion, but the low quality of services has kept "rural" a worse-off version of "urban".

If the building of infrastructure were the state's priority, providing social development education, health, housing, nutritious food, and decent employment would be the way to transform rural.

References

Adams, J. (1994). Economy as instituted process: Change, transformation, and progress. *Journal of Economic Issues*, *XXVIII*(2), June 1994.

Agarwal, B. (1981). *Agricultural modernisation and third world women: Pointers from the literature and empirical analysis*. ILO Working Paper.

Agarwal, B. (1994). *A field of one's own—Gender and land rights in South Asia*. Cambridge: Cambridge University Press.

Ahmed, N., & Nauriyal, D. K. (2023). Occupational and educational mobility among Indian Muslims: Primary survey-based evidence. *Millennial Asia*, *14*(2), 228–259. https://doi.org/10.1177/09763996211044376

Akram-Lodhi, A. H., & Kay, C. (2010). Surveying the agrarian question (part 1): unearthing foundations, exploring diversity, *The Journal of Peasant Studies*, *37*(1): 177–202. DOI: 10.1080/03066150903498838

Amin, S. (Ed.). (1974). *Introduction to modern migration in West Africa*. London: Oxford University Press.

Anand, I., & Thampi, A. (2016). Recent trends in wealth inequality in India. *Economic and Political Weekly, 51*(50), 59–67.

Annual Status of Education Report (Rural). (2022). New Delhi: PRATHAM.

Athreya, V. B. (1983). Review of dynamics of rural transformation by C T Kurien. *Social Scientist*, *11*(10), 57–68.

Athreya, V. B., Djurfelt, G., & Lindberg, S. (1990). Barriers Broken: Production Relations and Agrarian Change in Tamil Nadu. New Delhi: Sage Publications.

Bagchi, Amiya Kumar (2004). Demographic Regimes and Social Reproduction of Labour. *Economic and Political Weekly 39*(30): 3367–3377.

Bailey, F. G. (1959). For a sociology of India? *Contributions to Indian Sociology, 1*(3), 88–101.

Bardhan, P. (1983). Labour tying in a poor Agrarian economy. *Quarterly Journal of Economics, 98*(3), 501–514.

Bardhan, P. (1984). Agrarian class formation. In P. Bardhan (Ed.), *Land, labour and rural poverty: Essays in development economics*. New Delhi: Oxford University Press.

Bathla, S., & Kumar, A. (2019). Factors contributing to income inequalities among agricultural households in India. *Economic and Political Weekly, 54*(21), 55–61.

Bernstein, H. (2004). 'Changing before our very eyes': Agrarian questions and the politics of land in capitalism today. *Journal of Agrarian Change*, *4*(1 & 2), 190–225.

Beteille, Andre (1996). Sociology and Common Sense. *Economic and Political Weekly*, *31*(35-36-37): 2361–2365.

Bhaduri, A. (2010). *Essays in the reconstruction of political economy*. New Delhi: Aakar Books.

Bhalla, G. S., & Chadha, G. K. (1983). *Green revolution and the small peasant: A study of income distribution among Punjabi cultivators*. New Delhi: Concept Publishing.

Binswanger-Mkhize, H. P. (2013). The stunted structural transformation of the Indian economy- agriculture, manufacturing and the rural non-farm sector. *Economic and Political Weekly, 48*(26 & 27), 5–13.

Block Fred. (2001). *Introduction to Polanyi, Karl the great transformation-the political and economic origins of our time*. Boston, MA: Beacon Press.

Braverman, A., & Stiglitz, J. (1982). Sharecropping and interlinking of Agrarian markets. *American Economic Review, 73*(4), 693–715.

Breman, J. (1991). A note on colonial state. In J. Breman & S. Mundle (Eds.), *Rural transformation in India*. New Delhi: Oxford University Press.

Breman, J., & Mundle, S. (Eds.). (1991). *Rural transformation in India*. New Delhi: Oxford University Press.

Burman, B. K. Roy, Singh, Y., Oommen, T. K., Joshi, P. C., Dube, S. C. (1974). Land Reforms in a Sociological Perspective. *India International Centre Quarterly, 1*(1): 51–68.

Byres, T. J. (1991). The Agrarian question and differing forms of capitalist Agrarian transition: An essay with reference to Asia. In J. Breman & S. Mundle (Eds.), *Rural transformation in India*. New Delhi: Oxford University Press.

Chakrabarti, S., & Kundu, A. (2009). Rural non-farm economy: A Nnote on the impact of crop-diversification and land-conversion in India. *Economic and Political Weekly*, *44*(12), 69–75.

Chakravarti, A. (2001). *Social power and everyday class relations-Agrarian transformation in North Bihar*. New Delhi: Sage Publications.

Chand, R., Lakshmi Prasanna, P. A., & Singh, A. (2011). Farm size and productivity: Understanding strengths of smallholders and improving their livelihoods. *Economic and Political Weekly, 46*(26 & 27), 5–11.

Chand, R., Srivastava, S. K., & Singh, J. (2017). Changes in rural economy of India, 1971 to 2012-lessons for job-led growth. *Economic and Political Weekly, 52*(52), 64–71.

Chandrashekhar, C. P., Jayati Ghosh & Anamitra Roychowdhury (2006). The 'Demographic Dividend' and Young India's Economic Future. *Economic and Political Weekly*, *41*(49): 5055–5064.

Chatterjee, S. (2021). Overcast minds: Caste consciousness and academic performance. *Economic and Political Weekly, 56*(13–18), 23–30.

Chatterjee, U., Murgai, R., & Rama, M. (2015). Employment outcomes along Rural-Urban gradation. *Economic and Political Weekly, 50*(26 & 27), 5–10.

Cohn, B. S. (1987). *An anthropologist among historians and other essays*. New Delhi: Oxford University Press.

Cohn, B. S. (1958). Changing tradition of a low caste. *The Journal of American Folklore, 71*(281), 413–421.

Cohn, B. S. (1959). Some notes on law and change in North India. *Economic Development and Cultural Change, 8*(1), 79–93.

Cohn, B. S. (1961). The pasts of an Indian village. *Comparative Studies in Society and History, 3*(3), 241–249.

Damodaran, Harish, (2024 a), More rural households now 'agricultural': what data says, what this means, https://indianexpress.com/article/explained/comeback-of-krishi-in-bharat, accessed on 12.11.2024

Damodaran, Harish, (2024 b), Explained: The paradox of stagnant rural wages, https://indianexpress.com/article/explained/explained-economics/stagnant-rural-wages accessed on 12.11.2024

Damodaran, Harish, (2024 c), Has agriculture performed better under Modi govt?https://indianexpress.com/article/explained/explained-economics/agriculture-status-modi-govt-9606866/ accessed on 12.11.2024

Dantawala, M. L. (1970). *From stagnation to growth: Relative roles of technology, economic policy and Agrarian institutions*. Presidential address, Indian Economic Association, Fifty-Third Conference, Gauhati, December 1970.

Das, Arindam, Niladri Sekhar Dhar, Kunal Munjal, & Madhura Swaminathan (2024). Patterns of Income Growth in an Eastern Uttar Pradesh Village, 2006–23. *Economic and Political Weekly, 59*(43), 54–61.

Das, Panchanan & Madhurima Saha (2022). Occupation, Earning and Gender- An Analysis with the Labour Force Survey in India. *Economic and Political Weekly, 57*(29), 115–122.

Deolalikar, A. B. (1981). The inverse relationship between productivity and farm size: A test using regional data from India. *American Journal of Agricultural Economics, 63*(2), 275–279.

Deshpande, R., & Palshikar, S. (2008). Occupational mobility: How much does caste matter? *Economic and Political Weekly, 43*(34), 61–70.

Dev, S. M. (2012). *Small farmers in India: Challenges and opportunities*. WP-2012-014. Mumbai: Indira Gandhi Institute of Development Research.

Dewey, John (1915), Democracy and Education (New York: Dover Publishing, 2004).

Dorin, B. (2017). India and Africa in the Global Agricultural System (1961–2050) towards new sociotechnical regime? *Economic and Political Weekly, 52*(25 & 26), 5–13.

Dumont, L., & Pocock, D. (1957). For a sociology of India. *Contributions to Indian Sociology, 1*(1), 7–72.

Djurfeldt, G., Athreya, V., Jayakumar, N., Lindberg, S., Rajagopal, A., & Vidyasagar, R. (2008). Agrarian change and social mobility in Tamil Nadu. *Economic and Political Weekly, 43*(45), 50–61.

Ghatak, M., & Roy, S. (2007). Land reform and agricultural productivity in India: A review of the evidence. *Oxford Review of Economic Policy, 23*(2), 251–269.

Ghose, A. K. (2023). India's exclusive growth. *Economic and Political Weekly, 58*(6), 35–42.

Goel, Vamika (2024). Trends and Determinants of Rural Non-farm Employment in India. *Economic and Political Weekly, 59*(43), 33–39.

Goodman, D., Sorj, B., & Wilkinson, J. (1987). *From farming to biotechnology: A theory of agro-industrial development*. Oxford: Basil Blackwell.

Guerin, I., Venkatasubramanian, G., & Kumar, S. (2015). Debt bondage and the tricks of capital. *Economic and Political Weekly, 50*(26 & 27), 11–18.

Harriss, J. (1991). Agriculture/Non-agriculture linkages and the diversification of rural economic activity: A South Indian case study. In J. Breman & S. Mundle (Eds.), *Rural transformation in India*. New Delhi: Oxford University Press.

Harriss, J., & Jeyarajan, J. (2016). Rural Tamil Nadu in the liberalization era- what do we learn from village studies? In P. J. Himanshu & G. Rodgers (Eds.), *The changing village in India-insights from longitudinal research*. New Delhi: Oxford University Press.

Himanshu (2024). Dynamics of Change in Rural India. *Economic and Political Weekly, 59*(43), 30–32.

Himanshu, B. J., & Lanjouw, P. (2016). Non-farm diversification, inequality and mobility in Palanpur. *Economic and Political Weekly, 51*(26 & 27), 43–51.

Himanshu, Budget Must Address Falling Incomes, Indian Express, 14 January 2023. Retrieved May 30, 2023, from https://indianexpress.com/article/opinion/columns/budget-must-address-falling-incomes-8378693/

Himanshu, P. J., & Rodgers, G. (Eds.). (2016). *The changing village in India-insights from longitudinal research*. New Delhi: Oxford University Press.

Hnatkovska, V., Lahiri, A., & Paul, S. B. (2013). Breaking the Caste Barrier: Intergenerational Mobility in India, *Journal of Human Resources*, *48*(2), 435–473.

Hunter, G. (1969). *Modernising peasant societies: A comparative study in Asia and Africa* (p. 81). London: Oxford University Press.

Islam, R. (1986). Non-farm employment in Rural Asia: Issues and evidence. In R. T. Shand (Ed.), *Off-farm employment in the development of Rural Asia*. Canberra: ANU, National Centre for Development Studies.

Iversen, V., Krishna, A., & Sen, K. (2017). Rags to Riches? Intergenerational occupational mobility in India. *Economic and Political Weekly, 52*(44), 107–114.

James, K. S. (2010). Demographic change, gender and health in India: Theory and evidence. In K. S. James, A. Pandey, D. W. Bansod, & L. Subaiya (Eds.), *Population, gender and health in India-methods, processes and policies*. New Delhi: Academic Foundation.

Jan, M. A., & Harriss-White, B. (2012). Three roles of the agricultural markets: A review of ideas about agricultural commodity markets in India. *Economic and Political Weekly, 47*(52), 39–52.

Jarvis, B. F., & Song, X. (2017). Rising intragenerational occupational mobility in the United States, 1969 to 2011. *American Sociological Review*, *82*(3), 568–599. https://doi.org/10.1177/0003122417706391

Jerath, A. J. (2013, July 21). In capitalism, you can buy a Merc and hire a Brahmin driver. *Times of India*. https://timesofindia.indiatimes.com/home/sunday-times/all-that-matters/In-capitalism-you-can-buy-a-Merc-and-hire-a-Brahmin-driver/articleshow/21204849.cms

Jha, P. (2022). Repeal of the 'black farm laws' A momentous battle won, but the war ahead is long and hard. *Economic and Political Weekly, 57*(5), 10–13.

Jodhka, S. S. (2016). Revisiting the rural in 21st century India. *Economic & Political Weekly*, *51*(26 & 27), 5–7.

Jodhka, S. S., & Simpson, E. (Eds.). (2019). *India's villages in the 21st century-revisits and revisions*. New Delhi: Oxford University Press.

Jodhka, S. (2002). Nation and village images of rural India in Gandhi, Nehru, and Ambedkar. *Economic and Political Weekly, 10*, 3343–3353.

Joshi, P. C. (2021). Agrarian Structure and the Rural Poor: Review of Perspectives. In Prashant K. Trivedi (Ed.), *Land and Labour in Indian Agriculture-Discourses on Growth and Equity* (pp. 107–120). New Delhi: Sage Publications.

Kulkarni, S., & Hatekar, N. (2013). Stereotypical Occupational Segregation and Gender Inequality: An Experimental Study. *Economic and Political Weekly*, *48*(32), 112–120.

Kumar, K. (2014). Rurality, modernity, and education. *Economic & Political Weekly, 49*(2), 38–43.

Kumar, R. (2019). India's green revolution and beyond-visioning Agrarian futures on selective readings of Agrarian past. *Economic and Political Weekly, 54*(34), 41–48.

Kumar, S. (2021). Economic transition to transformation in India. In R. S. Ghuman (Ed.), *Towards peace, harmony and happiness: Transition to transformation*. Chandigarh: Centre for Research in Rural and Industrial Development.

Kumar, S., Heath, A., & Heath, O. (2002). Determinants of Social Mobility in India, *Economic & Political Weekly*, *37*(30), 2983–2987.

Kundu, R. K., & Chattopadhyay, A. K. (2018). Spatial-temporal variations of crop diversification: A block level study in West Bengal. *Economic and Political Weekly, 53*(21), 58–64.

Kurien, C. T. (1983). *Dynamics of rural transformation: A case study of Tamil Nadu-1950–1980*. New Delhi: Orient Longman.

Lerche, J. (2013). The Agrarian question in neoliberal India: Agrarian transition bypassed? *Journal of Agrarian Change*, *13*(3), 382–404.

Lerche, J., Shah, A., & Harriss-White, B. (2013). Introduction: Agrarian question and left politics in India. *Journal of Agrarian Change*, *13*(3), 337–350.

Lerche, J., & Shah, A. (2018). Conjugated oppression within contemporary capitalism: Class, caste, tribe and agrarian change in India. *Journal of Peasant Studies*, *45*(5–6), 927–949.

Lewis, O., & Barnouw, V. (1956). Caste and the Jajmani system in a North Indian village. *The Scientific Monthly*, *83*(2), 66–81.

Lieten, G. K., & Srivastava, R. (1999). *Unequal partners- power relations, devolution and development in Uttar Pradesh*. New Delhi: Sage Publications and Indo-Dutch Studies on Development Alternatives.

Lobo, C., Chattopadhyay, N., & Rao, K. V. (2017). Making smallholder agriculture climate smart. *Economic and Political Weekly, 52*(1), 53–58.

Mahajan, K., & Nagaraj, R. (2017). Rural construction employment boom during 2000-12-evidence from NSSO surveys. *Economic and Political Weekly, 52*(52), 54–63.

Majumder, R. (2010). Inter-generational Mobility in Educational and Occupational Attainment: A Comparative Study of Social Classes in India, *Margin: The Journal of Applied Economic Research*, *4*(4), 463–494.

Majumdar, D. N. M. C. Pradhan, C. Sen & S. Misra Inter-Caste Relations in Gohanakallan- A Village near Lucknow in D. N. Majudar (Ed.). (1955). *Rural profiles*. Lucknow: The Ethnographic & Folk Culture Society, U.P.

Mann, R. S. (2017). Cropping pattern in Punjab (1966–67 to 2014–15). *Economic and Political Weekly*, *52*(3), 30–33.

Marriott, M. (1955). Little communities in an Indigenous civilisation. In M. Marriott (Ed.), *Village India: Studies in little community* (pp. 171–222). Chicago: University of Chicago Press.

Martin, N. (2015). Rural elites and limits of scheduled castes assertiveness in Rural Malwa, Punjab. *Economic and Political Weekly, 50*(52), 37–44.

Mayer, A. C. (1960). *Caste and kinship in Central India: A village and its region*. London: Routledge and Kegan Paul.

Mehta, B. S., & Shree, M. (2017). Inequality, gender, and socio-religious groups. *Economic and Political Weekly, 52*(8), 56–60.

Mellor, J. W. (1968). *Developing rural India: Plan and practice* (p. 358). New York: Cornell University Press.

Mensah, J. T., Adu, G., Amoah, A., Abrokwa, K. K., & Adu, J. (2016). What drives structural transformation in Sub-Saharan Africa? *African Development Review*, *28*(2), 157–169.

Mishra, S., & Singh, K. (2021). Scale neutrality in Indian agriculture. *Economic and Political Weekly, 56*(26 & 27), 7–13.

MoEFCC (2015). First Biennial Update Report to the United Nations Framework Convention on Climate Change, New Delhi: Ministry of Environment, Forests and Climate Change.

Mohanty, B. B. (2021). Agricultural Modernisation in Rural Maharashtra: Myth and Reality. In Trivedi, Prashant K. (Ed.), *Land and Labour in Indian Agriculture-Discourses on Growth and Equity* (pp. 168–187). New Delhi: Sage Publications.

Moore Jr, Barrington (1967). Social Origins of Dictatorship and Democracy: Lord and Peasant in Making of Modern World. London: The Penguin Press.

Mukhopadhyay, S., & Lim, C. P. (1985). *Development and diversification of rural industries in Asia*. Kuala Lumpur: Asian and Pacific Development Centre.

Myrdal, G. (1968). *Asian drama—an enquiry into the poverty of nations*. New Delhi: Kalyani Publishers.
Nambissan, G. B. (2009). *Exclusion and discrimination in schools: Experiences of Dalit children, working paper series children, social exclusion and development* (Vol. 1, Issue 1). Indian Institute of Dalit Studies. New Delhi: UNICEF.
Nambissan, G. B. (1996). Equity in education: Schooling of Dalit children in India. *Economic and Political Weekly, 31*(16–17), 1011–1024.
Nambissan, G. B. (2012). Private schools for the poor-business as usual? *Economic and Political Weekly, 47*(41), 51–58.
Namboodiripad, E. M. S. (1984). The marxist theory of ground rent: Relevance of the study of Agrarian question in India. *Social Scientist, 12*(2), 3–15.
Nussbaum, M. C. (2008). *Education for profit, education for freedom*. Special Lecture 1. Kolkata: Institute of Development Studies.
Opler, M. E., & Singh, R. D. (1952). Two villages in Eastern Uttar Pradesh (U.P.), India: An analysis of similarities and differences. *American Anthropologist, 54*(2 part 1), 179–190.
Opler, M. E. (1956). The extensions of an Indian Village. *The Journal of Asian Studies, 16*(1), 5–10.
Otten, T., & Simpson, E. (2016). F G Bailey's bisipara revisited. *Economic and Political Weekly, 51*(26 & 27), 25–32.
Pandey, A. (2023). Explaining the U-curve trend of female labour force participation in Bihar. *Economic and Political Weekly, 58*(30), 39–48.
Pandey, L., & Amarender Reddy, A. (2012). Farm productivity and rural poverty in Uttar Pradesh: A regional perspective. *Agricultural Economics Research Review, 25*(1), 25–35.
Patnaik, U. (1987). *Peasant class differentiation—a study in method with reference to Haryana*. New Delhi: Oxford University Press.
Patnaik, U. (2007). New data on arrested development of capitalism in Indian agriculture. *Social Scientist, 35*(7–8), 4–23.
Patnaik, U., & Patnaik, P. (2016). *A theory of imperialism*. New Delhi: Tulika Books.
Pawde, B., Shaw, T. S., & Trivedi, P. (2022). Household consumption expenditure inequality in Rural India (1993–94 to 2011–12). *Economic and Political Weekly, 57*(11), 40–49.
Piketty, T. (2014). *Capital in the twenty-first century*. Cambridge, MA, & London, England: The Belknap Press of Harvard University Press.
Polanyi, K. (1944) (2001 edition). *The great transformation- The political and economic origins of our time*. Boston: Beacon Press.
Raj, K. N. (1969). Some questions concerning growth, transformation and planning of agriculture in developing countries. *Journal of Development Planning, 1*, 15–38.
Raj, K. N. (1970). Ownership and Distribution of Land. *Indian Economic Review, 5*(1), 1–37.
Rajagopalan, C. (1961). The Rural-Urban Continuum: A Critical Evaluation. *Sociological Bulletin, 10*(1), 61–74.
Ramachandran, V. K. (2011). The state of Agrarian relations in India. *The Marxist, XXVII*(1–2), 51–80.
Rao, C. H. H. (1965). *Agricultural production functions, costs and returns in India*. New Delhi: Asia Publishing House.
Rao, C. H. H. (2022). Rising inequalities in income in India-the key role of sociopolitical factors. *Economic and Political Weekly, 57*(51), 29–36.
Rawal, V. (2013). Changes in the distribution of operational land holdings in India: A study of sample survey data. *Review of Agrarian Studies, 3*(2), 73–104.

Rawal, V., & Saha, P. (2015). Women's Employment in India: What do Recent NSS Surveys of Employment and Unemployment Show? SSER Monograph 15/1. New Delhi: Society for Social and Economic Reserach.

Rizvi, S. N. R. (1988). *Atharhwin Sadi ke Zamindar* (*Eighteenth Century Landlords*). New Delhi: People's Publishing House.

Sahasranaman, A., & Kumar, N. (2022). Income and inequality across Rural-Urban. Occupational and caste divides. *Economic and Political Weekly, 57*(26 & 27), 23–30.

Sanyal, K. (2007). *Rethinking capitalist development: Primitive accumulation.* Governmentality, and Post-Colonial Capitalism. New Delhi: Routledge.

Santolo, Mary Di, Isabelle Guérin, Sébastien Michiels, Cécile Mouchel, Arnaud Natal, Christophe Jalil Nordman, Venkatasubramanian, G. (2024). A Decade in Rural Tamil NaduSocio-economic, Labour, and Migration Trends from an Original Longitudinal Household Survey. *Economic and Political Weekly 59*(43), 62–71.

Satheesha, B., & Neha Bailwal (2024). COVID-19 and Structural Change, *Economic and Political Weekly, 59*(43), 40–47.

Sbriccoli, T. (2016). Land, labour and power-A Malwa Village, 1954–2012. *Economic and Political Weekly, 51*(26 & 27), 8–16.

Scaria, S. (2010). Changes in land relations: The political economy of land reforms in a Kerala Village. *Economic and Political Weekly, 45*(26 & 27), 191–198.

Sen, A. K. (1964). Size of holdings and productivity. *Economic & Political Weekly, 16*(5–7), 323–326.

Sen, A. K. (1985). *Women, technology and sexual division.* New York: UNCTAD, Geneva and UNIRTIAW.

Sen, S., & Raju, S. (2006). Globalisation and expanding markets for cut-flower: Who benefits? *Economic & Political Weekly*, *41*(26), 2725–2731.

Shah, A., & Harriss-White, B. (2011). Resurrecting scholarship on Agrarian transformations. *Economic and Political Weekly*, *46*(39), 13–18.

Shah, G., Harsh Mander, Sukhdeo Thorat, Satish Deshpande and Amit Baviskar (2006). Untouchability in Rural India, New Delhi: Sage Publications.

Sharma, S. P. (1969). Indian village as unit of study-I. *Economic and Political Weekly, 4*(33), 1347–1354.

Siddiqui, M. Z., Lahiri-Dutt, K., Lockie, S., & Pritchard, B. (2017). Reconsidering women's work in Rural India- analysis of NSSO data, 2004–05 and 2011–12. *Economic and Political Weekly*, *52*(52), 45–51.

Singh, Rahul Kumar & Sarthak Gaurav (2024). Evolution of Social Networks in Kanzara Village Changing Patterns of Social, Informational, Agricultural, and Financial Interactions. *Economic and Political Weekly, 59*(43), 72–79.

Singh, K. N. (1968). The territorial basis of medieval town and village settlement in Eastern Uttar Pradesh, India. *Annals of the Association of American Geographers*, *58*(2), 203–220.

Singh, P., Goyal, M., & Choudhary, B. B. (2021). Drivers of foodgrain productivity in Uttar Pradesh- panel data analysis. *Economic and Political Weekly*, *56*(38), 40–45.

Singh, R. D. (1956). The unity of an Indian Village. *The Journal of Asian Studies*, *16*(1), 10–19.

Singh, S. K. (2016). De-ploughing the 'Rural'-Urban myths about rural consumption. *Economic and Political Weekly*, *51*(25). http://www.epw.in/journal/2016/25/web-exclusives/de-ploughing-%E2%80%9Crural.html

Singh, S. B. (2016). Living together separately: Dalits in emergent collectivity in a Rajasthan Village. *Economic and Political Weekly, 51*(33), 65–72.

Soundararajan, V., Sharma, G., & Bapuji, H. (2023). Caste, social capital and precarity of labour market intermediaries: The case of Dalit labour contractors in India. *Organization Studies*, 0(ja). https://doi.org/10.1177/01708406231175319

Srinivas, M. N. (1951). The social structure of a Mysore village. *The Economic Weekly, 3*(42–43), 1051–156.

Srinivas, M. N. (1955). *India's villages.* Calcutta [Kolkata]: West Bengal Government Press.

Srivastava, R. (2016). Assessing change; Land, labour and employment in an Eastern UP Village, 1994–2012. In P. J. Himanshu & G. Rodgers (Eds.), *The changing villages in India-insights from longitudinal research.* New Delhi: Oxford University Press.

Stiglitz, J. (2001). *Foreword to Polanyi, Karl the great transformation-the political and economic origins of our time.* Boston: Beacon Press.

Tilche, A. (2016). Migration, bachelorhood and discontent among patidars. *Economic and Political Weekly*, *51*(26 & 27), 17–24.

Thorner, A. (1982). Semi-feudalism or capitalism? Contemporary debates on classes and modes of production in India, *Economic and Political Weekly*, *17*(51), 2061–2066.

Thorat, S., & Dubey, A. (2012). Has Growth Been Socially Inclusive during 1993–94-2009–10? *Economic & Political Weekly*, *47*(10), 43–54.

Tripathi, A. K., & Sindhi, S. (2020). Droughts, heatwaves and agricultural adaptation-A historical account for India. *Economic and Political Weekly, 55*(26 & 27), 5–12.

Trivedi, P. K. (2011). Changing Agrarian scene: Sociology of land, wages, and indebtedness. In M. Mohanty (Ed.), *India social development report-2010-the land and the marginalized* (pp. 243–256). New Delhi: Oxford University Press.

Trivedi, P. K. (2022). Reinforcing exclusions: Caste, patriarchy and land reforms in India. *Journal of Land and Rural Studies, 10*(2), 262–277.

Trivedi, P. K., Ali, M., & Satpal. (2022). Farmer producer organisations in North India: Potentials and challenges. *International Journal of Rural Management*, 09730052221107730.

Unni, J. (2001). *Earnings and education among ethnic groups in Rural India.* Working Paper No. 124. New Delhi: Gujarat Institute of Development Research.

Veluguri, D., Ramanjaneyulu, G. V., & Jaacks, L. J. (2019). Statewise report cards on ecological sustainability of agriculture in India. *Economic and Political Weekly, 47*(54), 19–27.

Vijayabaskar, M., Narayanan, S., & Srinivasan, S. (2018). Agricultural revival and reaping the youth dividend. *Economic and Political Weekly, 53*(26 & 27), 8–16.

World Bank (2007). India- Land Policies for Growth and Poverty Reduction. Report Number 38298-IN. July 9. Agriculture and Rural Development Sector Unit. India Country Management Unit. South Asia Region.

Index

2020 Rural Transformation Survey 103

Adams, J. 4
Africa 139
Agarwal, B. 132, 137
agrarian economy 90, 119–120, 139, 151, 163
agricultural activities 164
agricultural assets 145
agricultural changes 120
agricultural economics 121
agricultural economy 91, 137
agricultural labour 89–90, 92, 147, 159; animal farming 147; non-OBC–SC–ST groupshare of workers 147; withdrawal of Dalit workers from 91
agricultural land 104
agricultural landownership 106, 154; stark inequality in 154
agricultural land tenancy 155
agricultural markets 161–163
agricultural occupations 93, 159
agricultural output 126–128
Agricultural Produce Marketing Act (APMC) 161
agricultural sector 138, 161, 169
agriculture 158; average productivity of paddy 158; climate change on 164; crop yield 128–130 ecological sustainability of 163; farm implements 159; greenhouse gases 163; Kharif season 158; labour relations in 136; policy orientation 159–160; productivity and shrinking diversity of crops 158; Rabi season 158; rent-seeking upper caste (UC) households 121; rural non-farm sector (RNF) 165; social groups in cultivation 159; technological adoption in 130–134 in Uttar Pradesh 120; wheat productivity 158
agrometeorological services 164
Ahirs 33, 38–39
Ambedkar, B. R. 9
animal farming 90, 92, 146–148; agricultural labour and 147; cultivation and 147; in primary occupation of Dalits 89
Asia 139
asset ownership 144–145
Athreya, V. B. 14

backward Muslims 153
Bailey, F. G. 11, 28
Begar (unpaid labour) 39
Bhagats 39
Binswanger-Mkhize, H. P. 151, 165
Brahmins 98–99
Breman, J. 118
Bundelkhand 70, 136, 154
Burman, B. K. 95
Byres, T. J. 118

capitalism 3, 5, 77
capitalist economy 8, 119
capitalist market 8
cash rent-based contracts 173; cheap and controlled labour force 94
caste 4; capital–caste dynamics 171; and capitalism 13, 78; and capital reinforce 13; caste location 77; conflicts 16; Dalit students 62; domination of landed upper jatis 20; in downward mobility 76; economic development 76–77; and gender 17; and gender discrimination 62; humiliation in everyday life 40; and joint family 10; landownership 41,

46, 98; landownership and political influence 15; and land possession 109–112; and mobility 76; service castes 38; significance of 16; in social organisation 29; upper castes' cultural ethos 79; upper castes (UC) 46; in upward mobility 76
caste communities 36
caste discrimination 62
caste domination 91
caste hierarchy 59
caste-land inequalities 162
Census of India 9
Chakrabarti, S. 130
Chamars 39–41
Chand, R. 126
chemical fertilisers 130
child marriages 38
China 52
Chitrakoot 52
civilisational values 28
climate change 164
climate compatibility 161
Cohn, B. S. 25, 34–36
colonial modernity 9
commercial agriculture 126
Constitution of India 35
consumerism 13
contemporary rural transformation 150–151, 155–156
cooperative movement, in India 162
Covid-2019 lockdown 8
cropping pattern 121–126, 122–123
culture 3

Dalit community 35; caste discrimination against 38
Dalits 15–16, 113; agricultural labourers 94; landownership possibilities 98; in salaried employment 89
Dantawala, M. L. 131
Das, Arindam 2
demographic changes 51; activity status 57–60; economic consequences of 51; educational status 62–65; in Gohanakallan 53, 57; in Seemli 53, 55, 57; in Senapur 53, 55, 57; social relations;166–168; women's negotiating power 51
demographic transition 51, 73, 143, 171; occupational mobility 143; policy agenda 52; population policies 52
Deshpande, R. 76–77
dobhi taluka 29
drainage system 36
Dumont, L. 28–29, 31

education 60–62
embeddedness, concept of 3
Essential Commodities Act 2020 161
Europe 139

fair and remunerative price (FRP) regime 125
family labour: backward classes and Dalits 136; elementary technology and 7; of farming households 119; self-exploitation of 126; and technology 127, 136
farmer-producer organisation (FPO) 162
Farmers (Empowerment and Protection) Agreement of Price Assurance and Farm Services Act 2020 161
Farmers' Produce, Trade, and Commerce (Promotion and Facilitation) Act 161
Farm Laws 161
feminisation, of agriculture 93
First UP Rural Transformation Survey 20, 22, 40, 152; aspects of rural life 20; employment and income sources 21; household income, expenditure, indebtedness, and living status 21; income-based inequalities 21; land and labour relations 20; landownership pattern 21; technology and productivity 20; village background information 20
food grain cultivation 127
food grain production 120
free market 161

Gandhi 9
Gaurav, S. 2
gender biases 78
gendered segregation 93
gender equality 52
gender justice 167
Ghatak, M. 96
Ghose, A. K. 68
gig economy 151
Global North 139
Global South 2, 25, 27, 139, 159–160, 162–163, 167
Gohanakallan 0, 22, 62, 64, 70–71, 80, 97–98, 106, 108–109, 113, 123, 127,

155–156; average productivity of paddy in 158; households purchased farmland 156; households purchased land in 112; Kharif season in 158; landless households 152; non-agricultural labour in 148; non-OBC–SC–ST group 155; primary occupation 146; sale of land 113; SC households in 154; sex ratio 57; share of landless households in 101; social group-wise composition 143; Thakurs in 114; total wheat output of 127; tractors and trolleys in 145; working-age group in 80
Goodman, D. 130
The Great Transformation (Karl Polanyi) 3

Harriss-White, B. 169
Hinduism 41
Homestead land market 113
household assets 20, 22, 145
households survey 23; in data analysis 23; eastern UP village 23; religious group-wise distribution 24; SC category 24; social group-wise distribution of 24
Hunter, G. 131

income distribution 12
Indian villages 11
industrialisation 5
inequalities 65, 96, 99; caste-land inequalities 162; gender-based inequalities 167; income inequalities 66; in ownership of assets 68–73; poverty reduction 66; rural and urban areas 67; in rural areas 66; on social outcomes 67; socioeconomic indicators 65
informal moneylenders 163
informal tenant farmers 163
irrigated lands 98

Jatav-Chamars 97
Jatavs 45
jatis 31, 98
Jaunpur village 23
Job Guarantee Act 43
Joshi, P. C. 95

Kumar, K. 60
Kumar, S. 76
Kundu, A. 130
Kurien, C. T. 2

labour force participation rate (LFPR) 1, 166
labour markets 12, 19, 78; female work participation in 78; gendered occupational segregation and discrimination in 78
labour relations 18
land-caste congruence 4
land market 113
landownership patterns 73
land reform policies 157
land reforms discourse 127
land relations 152–158, 161; and caste 101–109; domination and subjugation 96; economic rationale, for land reforms 96; economistic considerations 95; historical legacy of 97–99; landlord–tenant-type old relations 96; land market and quality of land 112–113 landownership structure 99–101; sociological perspective on 95; sociology of 96; women and Dalits 96
Latin America 139
Lerche, J. 169
Lucknow 52

macroeconomy 75
Majumdar, D. N. 25, 29–30, 38, 41
Majumdar village 29
Malwa village 11
Marathas 77
market channels 135
market domain 162
market economy 7–8, 13, 17–18, 26, 77, 119, 144, 149–150, 169
market forces 4, 142, 157; caste 4
market integration 26, 123, 172
market intervention 7
market penetration 68
market price 163
market transactions 173
Martin, N. 15
Mayer, A. C. 11
Mellor, J. W. 6, 131
migrant Dalits 11
migration 9; circular migration of workers 19; migrant labour 150–151; rural–urban migration 140
minimum support price (MSP) 125, 163
mobility 76; among Dalits and Adivasis 77
modern education 60–62
Mundle, S. 118

Nambissan, G. B. 62
National Sample Survey Office (NSSO) data 12
Nehru, Jawaharlal 9
neoliberal globalisation 2, 65
neoliberalism 139, 160
non-agricultural economy 89
non-agricultural labour 80, 142, 145, 147
non-agricultural occupations 91
non-agricultural sector 118, 120, 142; self-employment in 170
non-farm sector 2
Noniyas 33, 97

occupational changes 76; among Dalits 76; class origin 76; gender 78; gendered change 86–90; primary and secondary occupations 84; social relations 166–168; in studied villages 80–84
occupational diversification 169, 171
occupational mobility 143, 146
Opler, M. E. 29, 33–34

Paddy 158
Palanpur 11
Palshikar, S. 76–77
Pandari 0, 42–44, 57, 64, 70–71, 98–99, 106, 109, 112–113, 115, 123, 127, 148, 155–156; average wheat productivity in 158; Brahmin-Thakur combined in 114; household activities 59; households purchased land in 112; illiterate population 144; landless households 152; non-OBC–SC–ST group in 155; SC and OBC social groups 143; share of landless households in 99; total wheat output of 127; tractors and trolleys in 145; working-age group in 80; working population in 59
pargana 29
Pasis 39, 41
Patnaik, P. 160
Patnaik, U. 160
person's identity 4
Pocock, D. 28–29, 31
Polanyi, K. 3, 161
political economy 13, 31, 78, 146
political institutions 3
productivity analysis 121
public policy 2, 139; on rural development 165–166

qualitative data 23, 145
quantitative data 22
quitting agriculture 142

Raj, K. N. 131
Ramachandran, V. K. 173
Rao, C. H. H. 132
rental market 145
restrictive labour laws 151
risk management 161
Roy, S. 96
rural credit markets 161, 163
rural economy 1, 118–119, 165; market penetration in 2; structural transformation 1
rural ecosystem 157
rural hinterlands 165
rural households 1
rural market 119
rural migrants 144, 151
rural non-farm enterprises (RNFs) 6–7, 140–141, 165; market economy 7
rural social hierarchy 96
rural societies 66
rural transformation 2, 60, 75, 95, 139; and agrarian change 118; agrarian sector 5; agricultural growth 6; cropping pattern 121–126 defined 2; education 60–62; farm and non-farm sectors 8; food production and employment generation 6; labour wages 165; land relations 152–158; *see also* land relations linear movement 17; market economy 17; "Sanskritisation" and "dominant caste" 13; society's developmental and cultural parameters 16; state, role of 139; Western-style industrialisation 139
rural–urban distinctions 141
rural–urban gradation 141
ryotwari system 98

Scaria, S. 16
Second Rural Transformation Survey 22–23, 40, 123, 133, 152
Seemli 0, 44–46, 71, 80, 98–99, 115, 125, 148; activity status pattern of 59; age-group-wise distribution 53; households purchased land in 112; landless households 152; Muslims population of 143; primary occupation 146; sex ratio 57; sugarcane output in 128; tractors and

trolleys in 145; wheat output of 127; working-age group in 80, 82
semi-feudalism 5
Sen, A. K. 126, 133, 137
Senapur 0, 30, 32–38, 53, 62, 64, 70–71, 79, 98, 103, 109, 114–115, 123, 127, 133–134, 143, 146, 148, 156; activity status 59–60 age-group-wise distribution 53; average wheat productivity in 158; Hindu population in 55; households purchased land in 112; Kharif season in 123, 158; landless households 152; landlessness among OBCs of 153; landlessness among SCs of 153; Muslims population of 143; non-OBC–SC–ST group in 155; purchased agricultural land 156; sale of land 113; SC households of 155; secondary occupation in 147; share of agricultural labour 146; share of holdings in 152; share of landless households in 99, 101; tractors and trolleys in 145; working-age group in 80, 82, 84; working-age males 86
Shah, A. 169
Sindhi, S. 164
Singh, P. 121, 126
Singh, R. K. 2
Singh, R. D. 33
Singh, S. K. 16
social change 12, 23, 60–61
social development 16, 25; of rural people 151; state intervention in 174; of working people 144
social equality 127
social formation 68, 78
social groups 77, 89, 145, 156; in cultivation 159; financial assets 145; households belonging 153; pre-existing inequalities 155; social group-wise disaggregated analysis 64; socioeconomic conditions of 149
social hierarchy 21
social inequalities 2, 139, 155; market dynamics 155
social milieu 78
social policy 15
social protection 151
social relations 137, 166–168
social security 119
social structure 96
Stiglitz, J. 3
structural transformation 78
sugarcane 125; cultivation 126
Swacch Bharat Abhiyan 38

technological adoption, in agriculture 130–134
tenancy contracts 173
Thakurs 33, 36, 98; customary respect to Brahmins 39; freedom fighters 36; modern education on 35
Tilche, A. 16
transformation 2, 17, 141–142, 145; *see also* rural transformation; in activity status 60; of agrarian relations 118; of agriculture sector 169, 173; characteristics 142; defined 2; demographic transition and occupational mobility 143; economic, political, and social forces 4; economic relations 74; gender-segregated labour 148; income and labour 172; of labour relations 173; macroeconomics 3; of rural areas 171; social relations 166, 166–168; socio-political transformation 22
Tripathi, A. K. 164

U-curve hypothesis 1
Unni, J. 61
upper castes (UC) 46
urban economy 8, 48, 150
urban employers 144
urban labour market 119, 150
Uttar Pradesh 17, 52, 155

village Brahmins 35
village community 10, 30, 150
villages 23, 28; caste networks 31; de-homogenisation of 31; in eastern Uttar Pradesh 30; economics, political economy, sociology, and political science 32; encroachment 11; functional and distinguishable social unit 29; gendered occupational structure 147–150; Gohanakallan 38; history, political economy and sociology 23; homogenous community 31; inter-caste relations 29; land and labour-surplus villages 30; land-deficient village 30; landownership structure 99–101; Pandari 42–44; Seemli 44–46; Senapur 32–38 ; shortage of

labourers 30; social change in 11; Tamil Nadu villages 15; trade and service establishments in 141; in western Uttar Pradesh 11
village society 10
village surveys 20

wealth inequality 67
welfare schemes 13
Western academia 19
Western societies 17
women 2; education 60–62; employment dynamics 12; employment opportunities 27; family's economic status 13; gender and economic development 13; in household chores 13; in low-productivity agrarian vocations 93; low-productivity economic activities 149; in low-productivity sectors 2; occupational distribution 88; traditional occupations 77; women workers 74; workforce participation 49
World Bank 157

Yadav castes 36, 41, 97–98

Zamindari abolition 39, 48
zamindari system 41
zamindars 30, 98; revenue system 98; and ryotwari areas 98

For Product Safety Concerns and Information please contact our EU
representative GPSR@taylorandfrancis.com
Taylor & Francis Verlag GmbH, Kaufingerstraße 24, 80331 München, Germany

www.ingramcontent.com/pod-product-compliance
Lightning Source LLC
LaVergne TN
LVHW021128300926
849160LV00010B/913
9781032961514